Theory, Design, and Applications of Unmanned Aerial Vehicles

Theory, Design, and Applications of Unmanned Aerial Vehicles

A.R. Jha, Ph.D.

CRC Press
Taylor & Francis Group
Boca Raton   London   New York

CRC Press is an imprint of the
Taylor & Francis Group, an **Informa** business

MATLAB® and Simulink® are trademarks of The MathWorks, Inc. and are used with permission. The Math-Works does not warrant the accuracy of the text or exercises in this book. This book's use or discussion of MATLAB® and Simulink® software or related products does not constitute endorsement or sponsorship by The MathWorks of a particular pedagogical approach or particular use of the MATLAB® and Simulink® software.

CRC Press
Taylor & Francis Group
6000 Broken Sound Parkway NW, Suite 300
Boca Raton, FL 33487-2742

© 2017 by Taylor & Francis Group, LLC
CRC Press is an imprint of Taylor & Francis Group, an Informa business

No claim to original U.S. Government works

Printed on acid-free paper
Version Date: 20160819

International Standard Book Number-13: 978-1-4987-1542-3 (Hardback)

Library of Congress Cataloging-in-Publication Data

Names: Jha, A. R., author.
Title: Theory, design, and applications of unmanned aerial vehicles / A. R. Jha.
Description: Boca Raton, FL : CRC Press / Taylor & Francis Group, [2016] |
Includes bibliographical references and index.
Identifiers: LCCN 2016019916 | ISBN 9781498715423
Subjects: LCSH: Drone aircraft--Design and construction.
Classification: LCC UG1242.D7 J53 2016 | DDC 623.74/69--dc23
LC record available at https://lccn.loc.gov/2016019916

Visit the Taylor & Francis Web site at
http://www.taylorandfrancis.com

and the CRC Press Web site at
http://www.crcpress.com

Printed and bound in the United States of America by
Edwards Brothers Malloy on sustainably sourced paper

Contents

Foreword

This book comes at a time when Western countries are facing multiple international crises such as bogus claims from China on the small islands in the South China Sea, India's northern state of Arunachal Pradesh, military occupation by China of the disputed Spratly Islands in South China that are claimed by six Asian countries, and violation of Vietnam drilling rights in the coastal region very close to Vietnam. The disputed region has deposits of oil, gas, and minerals where China has illegally occupied, leading to a brief military conflict. It should be noted that Communist China had illegally occupied Tibet in 1954 because the region has unlimited deposits of rare earth materials that are widely used in motors and generators, electric and hybrid electric automobiles, military jet engines, commercial jet transports, and various medical devices and equipment. In addition, the Middle East refugee crisis poses serious security and financial crises particularly for the United States and European countries due to unstable political and volatile situations in Afghanistan, Pakistan, Syria, and other Arab countries. According to the latest international news, communist China has illegally developed landing strips in the South China Sea closer to the Philippines for its jet fighters and bombers in spite of objections by the United States, UN, and Southeast Asian nations. China has the worst track record for military conflicts with its neighbors including Russia, Vietnam, India, and other small countries like

Philippines. Because of the likelihood of military conflict between China and Southeast Asian nations, the United States and European countries might be involved due to security agreements with some Asian nations. There is a great danger in dealing with China because of its aggressive and assertive behavior. Furthermore, Russian military aggression in Syria and Ukraine poses a serious threat to North Atlantic Treaty Organization (NATO) countries. I am extremely impressed with the author's justification for the development of UAVs in order to maintain international peace and security in the troubled regions of this world.

The author has summarized the potential performance capabilities and unique aerodynamic features of the UAV unmatched by any vehicle to this date. Its performance capabilities and unique design features have been discussed in seven distinguished chapters, each dealing with a specific topic. This particular aircraft offers quick reaction military capabilities with minimum cost and complexity and no loss of pilot. This vehicle requires minimum maintenance and a small support group. Comprehensive studies performed by the author indicate that autonomous capability is possible using high-speed computers, the latest and efficient algorithms, state-of-the-art RF and optical sensors, and high-resolution side-looking radar (SAR) for precision target tracking in the extended forward-looking sector. The author has described briefly the aerodynamic design concepts for significant reduction in RF and IR signatures to avoid enemy radar detection and IR missile attack. Comprehensive studies performed by the author confirm that the latest RF and EO sensors aboard the UAV will provide optimum aircraft safety and survivability while executing the assigned combat missions. Furthermore, the unique aerodynamic design features described herein will provide optimum vehicle performance and survivability with no danger to the platform and the sophisticated EO and IR sensors aboard the vehicle.

The author has performed several trade-off studies to evaluate techniques for significant reduction in RF and IR signatures of the vehicle and the pod-mounted RF and IR sensors. Additive manufacturing technology and the three-dimensional printing concept have been evaluated by the author for the development of pod-mounted sensors and munitions to achieve minimum weight, size, and RF signature. As far as reduction of IR signature is concerned, the author has proposed a unique concept for the exit of hot exhaust gases from jet engines with no compromise to engine performance. It should be noted that the proposed design concepts for hot exhaust gases for the engine have no impact on the cost and endurance of the vehicle.

The author placed heavy emphasis on vehicle survivability and safety, particularly when the aircraft performs as a hunter–killer UAV and is required to penetrate heavily defended enemy territory. The author has maintained perfect balance between the number of weapons carried aboard the aircraft and the endurance requirements under assigned military missions. Any imbalance would impact not only the vehicle endurance capability but also the UAV ability to complete the assigned mission in the allocated time frame. The author feels that deployment of smart materials with optimum emissivity, direction of hot exhaust gases, reduction of engine thrust without affecting the engine performance, and optimum temperature control of the hot exhaust gases may lead to significant reduction in the IR signatures of the UAV. IR signature studies undertaken by the author indicate that small IR signatures are contributed by the fuselage surfaces, control surfaces, and skin temperatures. However, the maximum IR signature is contributed by the jet engine hot exhaust gases. Essentially, the IR signature is strictly a function of percentage of the total IR energy over a specific IR spectral bandwidth, which is dependent on the engine tail pipe temperature. The studies further confirm that the IR signature will be maximum during afterburner operations and minimum under cruise conditions. This means that to minimize the IR signature, the UAV aircraft must be operated under cruise vehicle speeds.

Stealth features of this vehicle have been given top priority to ensure optimum survivability and safety of the UAV platform. The UAV is equipped with high-resolution forward-looking radar, forward-looking infrared (FLIR), side-looking radar (SAR) with precision tracking capability, EO sensors aboard the vehicle, and podmounted laser-based compact missiles such as Hellfire missiles with low RCSs. The vehicle is fully equipped with compact high-resolution scopes, RF data links with optimum margins, and GPS receiver and antennas to communicate with ground-based operators if necessary.

In order to reduce the weight and size of the onboard sensors, the author has suggested the use of advanced technologies. The author has recommended using INS/GPS as a central unit of the navigational complex. Note that the core of the INS/GPS is a strap-down inertial navigation system (SINS), which is implemented as MEMS and integrated with GPS or other space-based satellite navigation system (SNS). The intended UAV flight requires the use of a Kalman filtering technique and a significant amount of central processing unit (CPU) time to obtain the correct flight parameters and the corrected angular orientation errors using magnetometers, accelerometers, and other measuring devices. In order to reduce the CPU time, the author has proposed the use of robust and adaptive filtering algorithms. Comprehensive error correction techniques can use a compensation circuit to correct the navigation parameters of the INS/GPS. External correction can be used for azimuth and vertical channels of the INS/GPS and the parameters of angular orientation. Note that the development of efficient algorithms for the extrapolation of SINS errors in the compensation filter network is necessary to achieve the required accuracy of autonomous functioning of the INS when the GPS signal is not available. The navigation parameters are the position and velocity components of the UAV.

Because autonomous requirement for this vehicle is the principal specification requirement, the author has treated the most critical areas in great details such as integration of the unified modules of aircraft onboard navigation and landing equipment for this small aircraft, modernization of steering servo-drive control system of autonomous vehicle based on experimental data available use of multiwave laser Doppler anemometer to measure instant wind speed calculation of UAV kinematical parameters at takeoff, prediction of INS error dynamics in the integrated INS/GPS to achieve higher navigation accuracy, and adaptive control system, which is an important element of the automatic control system. The author has briefly described the importance of the Kalman filter and its modified versions to achieve a robust and adaptive filtering technology, which is best suited for the correction of the angular orientation errors.

It is interesting to mention that Dr. A.R. Jha has a distinguished track record of distilling the complex physical concepts into an understandable technical framework that can be extended to a practical application such as the UAV best suited for military and industrial segments. Furthermore, his big picture approach, which does not compromise the basic underlying science, is particularly remarkable as well as refreshing. His approach should help defense scientists and military planners to understand the potential benefits of UAV technology in undertaking dangerous battlefield tasks with no loss of pilots and equipment.

This book is well organized and the author provides the mathematical expressions to estimate the critical navigation error limits. The author clearly identifies the cost-effective features of the design approach with particular emphasis on the reliability and safety of the RF and EO sensors and survivability of the vehicle under rough weather conditions. The author describes the aerodynamic parameter limits for the stability of the vehicle under windy conditions.

I strongly recommend this book to a broad audience including research scientists, military planners, and graduate students who wish to enhance their knowledge for the next generation of UAVs and project directors involved in the design and review of combat-related vehicles.

Dr. A.K. Sinha
Sr. Vice President (retired)
Applied Materials, Inc.
Santa Clara, California

Preface

The principal objective of this book is to develop a cost-effective design and evaluate an unmanned autonomous air vehicle (UAAV) for specific military applications. Essentially, the UAAV aircraft is a pilotless aircraft and is fully equipped with high-speed computers and the latest highly efficient algorithms. It should be noted that the book merely describes the conceptual design of the aircraft with no pilot in the cockpit. This means all pilot functions are carried out by the onboard computers, algorithms, and electro-optical (EO) and electromagnetic sensors and components. It is important to mention that the initial test and evaluation functions could be carried out by a pilot in the cockpit who can monitor the performance parameters of the sensors and the vehicle itself. Important performance parameters of the sensors and the aerodynamic performance of the aircraft can be effectively monitored by the human pilot in the cockpit.

This book comes at a time when the security and well-being of the Western industrial nations are threatened by third-world Islamic radicals. In addition, the twenty-first-century global economy of the free nations depends strictly on the quality and depth of the technological innovations as well as international peace and security of the nations. Recent terrorist attacks by Islamic radicals in France and California have created a mass-scale refugee problem. It is critically important to point out that appropriate military actions undertaken by such unmanned aerial vehicles (UAVs) are necessary to maintain peace and security around the world. Since the subject is very complex and technologically challenging, the author has made multiple trips to university libraries to review the latest unmanned autonomous design techniques.

For the benefit of the reader, the author has described important security aspects of the UAVs in seven distinct chapters, each emphasizing its own importance. Mathematical equations for the aerodynamic quantities related to UAV performance have been provided with particular emphasis on critical vehicle automatic flight control system (AFCS) performance parameters. Note that each chapter offers a detailed description on the subject related to the chapter heading. It is important to mention that the UAV will provide the basic capability including the information, surveillance, and reconnaissance functions over the regions of interest. This vehicle has potential applications in border patrol to keep drug dealers and illegal immigrants away from the country borders. The vehicle offers meaningful military actions in hostile regions.

Chapter 1 identifies potential commercial and military applications of these vehicles. Combat capabilities of UAVs have been discussed in great detail in this chapter. This chapter also describes the drone structural aspects suitable for intelligence gathering, surveillance, and reconnaissance (ISR) missions. These vehicles offer excellent sources for the demonstration of geographic capabilities, and cooperative forest fire surveillance functions using a team of small UAVs have been discussed.

Chapter 2 describes UAV configuration exclusively for complex military applications. Hunter–killer vehicles are recognized as essential aircraft and are described with particular emphasis on their target hunting and killing capabilities. This chapter describes unmanned demonstrator aircraft suitable for maritime surveillance functions. A queuing model for the supervisory control for UAVs has been discussed with emphasis on onboard sensor requirements and critical aerodynamic configuration requirements.

Chapter 3 defines the performance requirements for the onboard EO, radio-frequency (RF), and critical electronic sensor and components. The author has performed trade-off studies on the accuracy requirements of the sensors and components to maintain normal aircraft performance with no compromise on the reliability and safety of the aircraft and its contents. The tracking, range accuracy, and reliability requirements for the laser, forward-looking infrared (FLIR) sensor, mini-side-looking radar, and forward-looking multipurpose radar have been summarized. For optimum survivability and safety of the vehicle, radar cross sections (RCSs) for the vehicle, all pod-mounted sensors, and missiles must be kept less than 0.001 m^2 to avoid enemy missile or radar attacks.

Chapter 4 concentrates on the UAV navigation system and AFCS requirements to make sure that automatic flight control laws are obeyed to maintain the vehicle flight traveling on the selected destination point. The author has placed maximum emphasis on flight safety and automatic control of the UAV. Electronic means for automatic error correction of the inertial navigation system (INS) have been specified. Efficiency requirements for the algorithms associated with the integrated inertial–satellite navigation system are summarized. Error prediction of INS error dynamics in the combined INS/GPS is briefly described. Fault detection and flight data measurement techniques are summarized with emphasis on the accuracy of the data measurement.

Chapter 5 focuses on the propulsion systems and electrical power requirements for the operation of sensors and weapons aboard the autonomous vehicle. The next generation of high-power batteries and fuel cells are described specially for ultraquiet UAVs to avoid deployment of the offensive weapons by the enemy. The author provides the latest information on a unique UAV fuel system currently being developed jointly by General Electric Company and Woodward Global Company of Colorado. This new fuel system will significantly reduce the fuel consumption and extend the UAV operational range capability.

Chapter 6 focuses on UAV technology. This is a very important and tough topic. The author had to conduct comprehensive research studies and collect the latest UAV technology material from the latest published papers available at the University of California, Irvine Science Library. Microelectromechanical system (MEMS)- and nanotechnology-based sensors and components for AFCS were evaluated for UAV applications. The latest technical papers on robotics were screened and evaluated for possible application for real-time, high-resolution simulation of autonomous vehicle dynamics.

Chapter 7 specifically deals with the survivability and safety of UAVs while operating in hostile regions. The author considers the stealth feature of the vehicle the most critical design requirement. Highly absorbent radar paints and materials are best suited to reduce RCS. Aircraft designers have achieved an RCS deduction of 100:1 using high-quality paint and specified radar-absorbing material of optimum thickness. Further, RCS reduction can be achieved by eliminating the sharp corners and edges that contribute to radar energy reflections. Besides the RCS problem, one has to face the infrared (IR) energy radiation from the high-temperature exhaust nozzle. IR reduction is very complex and requires a sophisticated solution. The author has dealt with the IR reduction technique in great detail in this chapter.

I thank Jessica Vakili, project coordinator at CRC Press, Taylor & Francis Group, who has been very patient in accommodating the last minute changes to the text. Last, but not the least, I thank my wife Urmila Jha and daughter Sarita Jha who inspired me to complete the book on time under the tight schedule. Finally, I wish to express my sincere thanks to my wife, who has been very patient throughout the preparation of this book.

MATLAB® and Simulink® are registered trademarks of The MathWorks, Inc. For product information, please contact:

The MathWorks, Inc.
3 Apple Hill Drive
Natick, MA, 01760-2098 USA
Tel: 508-647-7000
Fax: 508-647-7001
E-mail: info@mathworks.com
Web: www.mathworks.com

1

HISTORICAL ASPECTS OF UNMANNED AERIAL VEHICLES*

Introduction

An unmanned aerial vehicle (UAV) is known by various names, such as a remotely piloted aircraft (RPA) or an unattended air system (UAS) or simply a drone. Essentially, a UAV is considered an aircraft without a human pilot. All aerodynamic functions can be controlled either by the onboard sensors or by a human operator in a ground control location or by the deployment of autonomous electronic and electro-optical systems. The most basic functions of a UAV include intelligence, reconnaissance, and surveillance (IRS). However, an unmanned combat air vehicle (UCAV) is supposed to meet combat-related functions in addition to IRS capabilities, such as target tracking and deployment of defensive and offensive weapon systems against targets. A UAV can be equipped with simple electronic and physical sensors such as a barometer, global positioning system (GPS) receiver, and altimeter device. Sophisticated UAVs can be equipped with photographic, television, infrared, and acoustic equipment, compact synthetic aperture radar (SAR), and light detection and ranging (LIDAR) laser along with radiation, chemical, and other special sensors to measure pertinent parameters to accomplish critical missions. Navigation and control sensors are of critical importance. Furthermore, the onboard sensors can be controlled by the ground-based operator, by preprogrammed sensors, or by automated remote operating mode. In case of unattended combat air vehicle (UCAV) mode, mission requirements can be changed by

---

* The last section of this chapter (pp. 24–43) is adapted from Casbeer, D.W. et al., Cooperative forest fire surveillance using a team of small unmanned air vehicles, *International Journal of Systems and Science*, 37(6), 351–360, 2006.

the ground operator. UAV design scientists believe that when using the onboard and ground-based equipment, the UAVs can perform a wide range of missions, such as intelligence gathering, surveillance, reconnaissance, aerial mapping, antiterrorist activities, and emergency operations with remarkable speed. Scientists further believe that the development of compact inertial navigation equipment, exotic software and algorithmic maintenance for equipment calibration, and filtering and rapid and accurate processing of navigational information will enable UAV operators to perform important tasks with great accuracy and speed. Electrical design engineers are deeply involved in specific development of onboard software and hardware of next generation of computer vision and pattern recognition for navigation and UAV orientation. When the needed sensors and equipment are fully developed and available, UAVs can be deployed to create highly accurate images of the mouths of rivers, coastlines, ports, and settlements in critical regions. Typical physical parameters of small UAVs for commercial applications can be summarized as follows.

Typical Physical Parameters of UAVs for Commercial Applications [1]

- *Takeoff weight*: 6–16 lb
- *Airframe weight*: 5–9 lb
- *Wing span*: 5–7 ft
- *Fuselage length*: 4–8 ft
- *UAV speed*: 20–30 mph
- *Payload*: 5–10 lb
- *Flight endurance*: 10–25 h
- *Rating of electric motor*: 1 kW or 1.35 HP (some UAVs use gasoline engine, while others use an electric motor)
- *Takeoff speed*: 15–20 mph
- *Landing speed*: 15–20 mph
- *Runway length*: 40–60 ft
- *Maximum climb speed*: 16 ft/s
- *Turn radius*: 35–50 ft
- *Flight altitude*: 50–6000 ft (max)

Various Categories of Unmanned Vehicles

There are various types of unmanned vehicles currently in operation for commercial and military applications. Unmanned vehicles capable of flying at higher altitudes for military applications can be easily recognized or characterized by the aerodynamic configuration of the vehicle and its operating sensors and their functions. UAVs are designed to meet specific commercial principal functions and to operate at low-to-medium altitudes. Unmanned vehicles specially designed and developed for military applications are equipped with electro-optical and electromagnetic sensors and electronic devices to provide safe and rapid delivery of military goods and services. As mentioned earlier, UAVs generally operate from low-to-medium altitudes not exceeding 6000 ft.

UAVs for Border Patrol Operations

UAVs for military or combat operations are designed and configured to carry out specific IRS missions. Stringent structural, reliability, and stealthy aspects are given priority in the design of combat-based UAVs. Note that UAVs for border patrol activities are not required to meet stringent structural and heavy weapon requirements. UAVs with conventional features are best suited to undertake IRS missions (Figure 1.1); UAVs with rotary-wing vehicle configurations (Figure 1.2) could also be used for border patrol operations. But unmanned underwater vehicles, as shown in Figure 1.3, are ideal for undertaking underwater listening, detection, reconnaissance, surveillance, and tracking of hostile targets operating in coastal regions and tracking of submerged underwater submarines.

UAVs such as MQ-1 Predator or MQ-9 Reaper are recognized as *hunter–killer* UAVs and are best for combat operations. These UAVs are fully equipped with advanced navigation and communications systems, electro-optical and electromagnetic sensors, and weapons needed to undertake the stated combat mission requirements. The control of these combat vehicles is handled either by a competent ground station operator or by an autonomous system comprising sophisticated high-speed computers.

Figure 1.1   Unmanned aerial vehicles with intelligence, reconnaissance, and surveillance missions.

Figure 1.2   Configuration of "Raven" RQ-IIB micro-UAV showing approximate locations of various sensors.

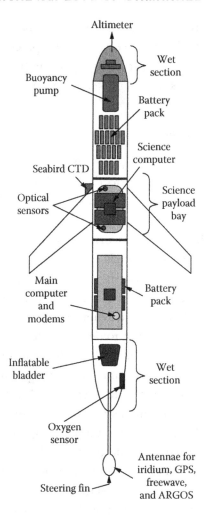

Figure 1.3 Long-range underwater drone.

Sensor requirements are very important for border inspection operations. Furthermore, authenticity in intelligence data collection missions and parameter accuracy in surveillance and reconnaissance missions are of critical importance, if successful mission accomplishment is the principal objective. Mission success is strictly dependent on the operator capability and the accuracy of the electro-optical sensor (IR cameras), electromagnetic system (SAR), high-speed signal processing equipment, and high-resolution display. Additional systems such as radio-frequency receiver with high sensitivity and light weapons such as Hellfire missiles for attack mission are essential equipment for border patrol activities. Patrol UAVs must be equipped with sensors capable of seeing through clouds, dust, smoke, and rain.

Chronological History of UAVs and Drones

Studies performed on the subject seem to indicate that the concept of UAVs or drones dates back to the mid-1800s, when Austria sent off an unmanned bomb-filled balloon as a tactical mean to attack Venice (Italy). The drones that are seen today are innovations of the early 1900s. Such drones were originally deployed for target practice to train military personnel. Consequently, they continued to be designed and developed during World War I, when the Dayton-Wright Airplane Company invented a pilotless aerial torpedo that would drop and explode at a particular preset time. Published articles reveal that the earliest attempt to demonstrate the UAV concept was made in 1943. Since then, the UAV or drone technology was developed for various commercial and military applications. As early as 1915 Nikola Tesla described the fleet of UAVs. The first RPV was developed in 1935 and demonstrated by a film. More UAVs were developed during World War II to train army gunners and to fly attack missions. Nazi Germany produced and deployed various versions of UAVs during World War II to destroy Allied fighters and bombers. German scientists and engineers were ordered by Hitler to design and develop miniaturized jet engines, which were used for the Teledyne Ryan Firebee in 1951. It is interesting to point out that Beechcraft Company also entered the game to develop UAVs for the U.S. Navy in 1955. However, they were more than remotely controlled [1] aircraft until the Vietnam conflict (1955–1967).

Two prominent UAV programs were initiated by the U.S. Department of Defense in the late 1950s or the early 1960s for advanced military missile applications. Some UAVs were specifically developed as per CIA specifications for covert programs. The CIA UAV programs were commissioned after September 11, 2001, and a major emphasis was placed in 2004 by CIA on intelligence gathering. According to published newspaper articles, the covert CIA program was deployed in mostly Islamic countries such as Pakistan,

Afghanistan, Yemen, and Somalia to collect intelligence data about Islamic radical and terrorists through the use of UAV loitering around the targets. The first CIA UAV program was known as Eagle program, and it was operated by the counterterrorism center to collect vital data on the terrorists. The first CIA UAV platform used the commercial off-the-shelf (COTS) components of garage door openers and model airplanes using COTS devices to realize substantial cost reduction. Note that useful performance data on COTS discrete components and circuits are readily available in *Popular Science* and *Mechanics Design* publications.

The USAF 160 Strategic Reconnaissance Wing had undertaken close to 3500 UAV missions during the Vietnam War. As a matter of fact, the official birth of U.S. UAVs, known as RPVs, began in 1959 when the U.S. officials were concerned about pilot losses and aircraft losses on hostile territory. This USAF thinking was more intensified when Francis Gary Powers and his U-2 reconnaissance plane were shot down over the Soviet Union in May 1960. A highly classified manned aerial vehicle program was launched under the code name "Red Wagon" [1]. Maximum deployment of UAVs was observed during the conflict between North Vietnam and U.S. naval combat aircraft in August 1964. As mentioned earlier, the CIA UAV program was commissioned after the September 11, 2001, terrorist attack on New York's World Trade Center. The intent of the CIA UAV program was the collection of intelligence on Vietnam territories and other hostile lands. Because of the dangers posed by Islamic radicals, major emphasis was placed on intelligence gathering on the terrorists. Furthermore, covert UAV activities were focused strictly on intelligence collection activities on terrorist groups operating particularly in countries like Pakistan, Yemen, and Somalia.

According to the Department of Defense, more than 5000 U.S. soldiers were killed and more than 1000 were missing or captured as prisoners of war. The USAF's 100th Strategic Reconnaissance Wing was deployed. Approximately 3425 UAV missions were carried out during the war, out of which 554 were lost. General Meyer, commander in chief of Strategic Air Command, stated that he is willing to take a risk to fly UAVs or drones at higher altitude, which will significantly save the lives of pilots. During the Yom Kippur War [4], the Soviet Union supplied surface-to-air missiles in Egypt and Syria was heavily damaged by the Israeli Air Force fighter jets without any pilot loss. This forced the Israeli Air Defense Command to rapidly develop combat-based UAVs with real-time surveillance capability. Real-time, high-resolution images by Israeli UAVs allowed the Israeli pilots to effectively penetrate the Syrian air defenses at the start of 1982 Lebanese War with no pilots down or killed. The proof of the concept of surprise agility was demonstrated by the Israeli pilots. Israeli pilots learned valuable experience by the combat flight simulators. Furthermore, in 1987 Israel pilots demonstrated stealth technology–based three-dimensional thrust vectoring flight, jet-steering UAVs.

The maturing and miniaturization technologies available during the 1980 and 1990 time frames demonstrated the greatest interest in UAV platforms by the U.S. defense authorities. In late 1990, the U.S. Department of Defense issued a UAV production contract for the joint development of the AAI Corp and Israeli Company Maziat. Many UAVs were used in the 1991 Gulf War. Department of Defense officials observed that the UAVs offer cheaper and best fighting machines with minimum cost and without any risk to aircrews. Later on, the Department of Defense scientists observed that UAVs equipped with Hellfire missiles and light ammunitions can replace manned fighters–bombers with minimum cost and zero risk to pilots who are actually sitting in the GCS, which is about thousands of miles from the scene of military action. To illustrate the strategic importance of UCAVs, the performance summary and capability of the General Atomics Predator M Q-9 platform is provided for the benefit of the readers.

This particular UCAV aircraft provides significantly improved surveillance and reconnaissance capabilities along with the best available offensive weapons such as air-to-ground Hellfire missiles. The performance capabilities of the MQ-9 vehicle were so impressive that about 50 countries or more are deploying this particular platform effective 2013.

Since 2008, the USAF has deployed 5331 UAV missions, which is twice the number of manned aerial vehicles. In the opinion of defense experts, the Predators equipped with Hellfire missiles are the most cost-effective UAVs because they are considered most effective in undertaking combat roles with no risk to the pilot or aircraft.

Furthermore, the UCAV missions can be terminated with no time if the military authorities would like so. Military experts feel that the Predators are best suited for orchestrating attacks by pointing precision laser beams on the targets. The laser beam is capable of putting a robot in a position to set off an attack on the target of interest with no collateral damage. Furthermore, the Predators can be operated via satellites by the ground-based pilots sitting in the GCS located at distances greater than 7500 mi or so from the scene of military action.

The U.S. Army's Global Hawk UCAV designated as RQ-4 is operated virtually autonomously using high-speed computers and sophisticated software. Such a system is considered a remote-controlled pilot or operator. The absence of a human pilot in the UCAV platform is the most outstanding benefit of autonomous control of the UCAV. It appears that the future generation of military authorities would follow this particular trend in military conflicts. It requires the user merely to hit the appropriate button "to take off and land," while the UCAV gets its directions via a GPS and reports back via a secured live feed. According to military experts, a Global Hawk has the operational capability to fly from the west coast of the United States and map out the entire state of Maine located on the east coast before having to return to its departure station. Some aerospace companies are engaged in the design and development of miniaturized UAVs, which can be launched from the soldier's hand and maneuvered through the streets at lower altitudes. An aerospace company in California has developed such miniaturized aerial vehicles. The Raven micro-air vehicle is considered the most versatile miniaturized UAV and is best suited for urban applications. The Raven can play critical roles especially in crowded unknown hostile lands. Counterterrorism studies performed by the author seem to indicate that Ravens are best suited for undertaking reconnaissance and surveillance missions in crowded urban environments to find insurgents, potential ambushers, and illegal immigrants. The studies further indicate that the miniaturized UAVs are most practical for surveillance and target tracking missions because they can fly at low and medium altitudes for days at a time. Note that insurgents or illegal immigrants hate to stay in the open for more than a few minutes for fear of UAVs locating them. Classification of the miniaturized UAVs is based on applications in a specific field.

The U.S. military authorities have provided the following miniature UAV classifications based on specific application:

- Remote sensing
- Domestic surveillance
- Oil, gas, and mineral exploitation
- Commercial aerial surveillance
- Policing activities by law enforcement agencies
- Forest fire detection
- Disaster relief during heavy rain, blizzards, snow, and flood conditions to reduce civilian casualties
- Armed attacks by terrorists or drug gangsters
- Precision search and rescue missions
- Maritime patrol
- Rescue operations in dangerous and difficult circumstances

UAVs Operated by Various Countries for Surveillance and Reconnaissance

Studies were undertaken by the author to estimate how many countries have UAVs in actual operations. Various UAV designers, aerospace companies, and Wikipedia published materials were reviewed with emphasis on the technical ability of various nations to design, develop, and operate UAVs. The following countries seem to possess UAVs mostly to conduct IRS missions. However, only a few countries such as the United States, France, the United Kingdom, and Israel have demonstrated the design, development, and operational capabilities of UCAVs. Furthermore, only the United States has demonstrated autonomous capability to operate UAV and UCAV platforms capable of deploying sophisticated electro-optical sensors, electromagnetic sensors, and missiles. The countries involved in possession of UAVs are as follows:

- United States
- United Kingdom
- France
- Israel
- Belgium
- Australia

- Germany
- Canada
- Japan
- China
- Russia
- Brazil
- South Africa
- India
- South Korea
- Ireland
- Vietnam
- Venezuela
- Republic of Congo

Comments

According to published documents, most of these countries have demonstrated surveillance and reconnaissance functions conducted by UAVs, but only a few countries have demonstrated the IRS and combat capabilities of unmanned combat aerial using sophisticated laser, radar, and offensive weapons such as Hellfire missiles. The author would like to make constructive comments about the Chinese "Guizhou" ("Soar Eagle Hawk") UAV. This particular UAV bears a distinct resemblance to Northrop Grumman "Global Hawk" UAV. Furthermore, it uses a Soviet-developed turbojet engine WP-13, which is widely used in MIG-21 jet fighters and SU-15 interceptors. It is very interesting to mention that 107 Indian Air Force MIG-21 fighters were crashed mostly because of mechanical reliability problems associated with the WP-13 power plant. Based on this comment, it seems uncertain whether the UAV design is actually Russian or Chinese.

Deployment Restriction on UAVs

Deployment of UAVs poses serious security problems for commercial flights irrespective of operating altitudes. In the United States, the Federal Aviation Agency (FAA) Modernization and Reform Act (MRA) of 2012 sets a deadline of September 30, 2015, for the agency to establish regulations to allow the use of commercial UAVs or drones.

FAA Designations and Legal Regulations

FAA considers the UAV as an aircraft system with no flight crew aboard the vehicle. This aircraft is called by various names such as a UAV, an RPV, an RPA, and a drone. Small or miniaturized or micro-aerial vehicles are flown only for recreation or fun and are operated under the voluntary safety regulations recommended by the Academy of Model Aeronautics. However, for other unmanned air operations in the United States, a Certificate of Authorization from the FAA is required to operate in national aerospace. The FAA MRA of 2012 has set up a deadline of September 30, 2015, for the agency to establish safety regulations for the commercial UAVs or drones. Furthermore, the agency states that it is illegal to operate a UAV without an official certificate issued by the agency. However, it approves noncommercial UAV flights under 400 ft if they comply with the Advisory Circular 91-57, Model Aircraft Operating Standards that were published in 1981.

Commercial unmanned aerial system (UAS) licenses were granted effective August 2013 on a case-to-case basis but subject to approval from the FAA. The FAA expects to unveil a regulated framework for UAVs weighing not more than 55 lb. Critical elements of a UAS are as follows:

1. Unmanned aircraft (UA)
2. GCS
3. Specialized data link or control line (SDL)
4. Related support equipment

For example, the reconnaissance platform RQ-7 Shadow UAS, consists of four elements, which include two GCS, one portable GCS, one launcher, two ground data terminals (GDTs), one portable GDT and one remote data terminal. For certain military applications or remote location operation, a maintenance support van is provided, which is equipped with emergency and necessary items.

Functional UAV Categories: A UAV falls into one of the five distinct functional categories as specified below:

1. *Reconnaissance UAV* is fully equipped to provide battlefield intelligence.

2. *Target UAV* is capable of providing ground and aerial gunnery functions of simulating a hostile aircraft or a laser-guided or semiactive radar missile.
3. *Combat UAV* provides attack capability, which is exclusively provided by UCAV.
4. *Logistic UAV* is designed to provide cargo and logistic supplies.
5. *Research and development* (*R and D*) activities provide further development of UAV technologies for possible integration in future UAV designs.

Functional Categories of UAS Based on Operating Range and Altitude (Estimated Values):

- *Handheld UAS* with range and altitude not exceeding 2000 ft.
- *Tactical UAS* with range close to 160 km and altitude not exceeding 18,000 ft.
- *Medium-altitude, long-endurance (MALE) UAS* has an operating range greater than 200 km and exceeding 30,000 ft.
- *Hypersonic UAS* has a range greater than 200 km and altitude greater than 50,000 ft.
- *Orbital UAS* travels in low-Earth orbit with range and altitude data not yet available.

UAV and UCAV Operations: According to published articles, the UAVs and UCAVs are being operated by U.S. Maritime Corp, USAF, and U.S. Navy. Specific details on various unmanned vehicles can be briefly summarized as follows:

- *Micro-UAV*: Its role is filled by BATMAV (WASP III design). The 1 lb microvehicle WASP III designed by a California company carries all the microsensors including a miniature altimeter, lightweight GPS-based navigation system, accelerometers, and a compact magnetometer that can be housed in a pod [5].
- *Low-altitude, long-endurance (LALE) UAV*: The role of this VAV is provided by the GNAT 750 system.
- *MALE UAV*: The role of this vehicle is provided by MQ-1 Predator or MQ-9 Reaper. Both of these UAVs are known as hunter–killer UAVs.

- *High-altitude, long-endurance (HALE) UAV:* The role of a UAV can be provided by the General Atomics Predator-9 UCAV. It can be stated that the military role of UAS vehicles can grow at unprecedented rates. Note that both the tactic-level and theater-level platforms have flown more than 100,000 flights in support of Operation Enduring Freedom and Operation IRAQ. HALE-designed UAVs have played impressive roles in combat. For example, General Atomics MQ-9 Reaper is considered a hunter–killer platform and has provided remarkable performance under combat environments. Note that the symbol "M" stands for multirole, "Q" stands for UAV, and number "9" indicates the series designation. It is necessary to mention that this UCAV is also known as Predator B and can perform as a hunter–killer platform. Both these platforms are known as HALE UCAVs. The MQ-9 Reaper offers the USAF a high-level, remotely piloted weapon platform capable of providing instant action and precise engagement [5]. In summary, it can be stated that the Reaper is a large derivation of the Predator series of UCAV and features more power in terms of both power plant and weapon delivery capability. Note that larger UAVs or UASs such as the Predator, the Reaper or the Global Hawk are always controlled by human operators sitting in GCSs, which are located thousands of miles away from the scene of action. Advancement in computer technology and artificial intelligence (AI) will allow the aircraft takeoff, landing, and flight control without a human operator in the loop. Note that AI systems are capable of making decisions and planning sequence of actions that will lead to the development of "fully autonomous" systems. As mentioned earlier, the UAV is a pilotless aerial vehicle, does not carry a human operator and uses aerodynamic forces to provide a vertical lift and to fly the vehicle autonomously. The UAV can carry a lethal or nonlethal payload. On the other hand, cruise missiles or other guided missiles are not considered UAVs because the vehicle itself is a weapon, even though it is also unmanned and in some cases remotely guided such as laser guidance.

Small Unmanned Aerial Vehicle

Military authorities believe that enormous success has been observed in various operations and the designers are now focusing on vehicle technology and materials. This allowed small UASs to be deployed on the battlefield with minimum drone losses. Research and development activities are now more focused on streamline designs of "drones" to undertake multiple missions, besides IRS missions. The roles of small UASs (SUAS) have expanded in other disciplines such as electronic attack strike missions, destruction of enemy air defense systems, and search and rescue missions under combat environments. Defense financial analysts believe that the estimated cost for a SUAV could vary from a few thousand dollars to tens of millions of dollars and their estimate weights could vary from 1–40,000 lb.

Civilian Applications of UAVs

So far, the majority discussion has been limited to military applications of UAVs and drones. Now, attempts will be made for civilian applications of UAVs.

Pizza Delivery by Small UAVs or Drones

Recently, some countries have made an attempt to deliver the pizza by drones. A Russian company pizza shop, Dodo, in Syktyvkar and a Bombay pizza shop in India successfully delivered *hot* pizzas within 30 min to their respective customers in Moscow (Russia) and in Bombay (India). The drones were monitored by GPS and video cameras to expedite the pizza delivery to the correct address. The drone flight was maintained at approximately 65 ft above ground level. Once the pizza arrives at the correct destination, the drone lowers the pizza through a cable and the customer at the prescribed destination picks up the pizza. As soon the customer picks up the pizza, the drone leaves for its base station.

Drone Deployments for Miscellaneous Commercial Applications

Comprehensive examination of commercial applications of drones reveals that drones could be deployed in various commercial applications including surveying of crops, aerial footage in the film-making industry, search and rescue operations in emergency situations, inspection of high-voltage power transmission lines and gas and oil pipelines, numerical count and tracking of wildlife, delivery of drugs and medical supplies to remote and inaccessible areas, keeping an eye on private properties, monitoring using drone-equipped video cameras by animal rights advocates to determine any illegal hunting in progress, spot-checking of illegal fox hunting, missions for forest fire detection, border patrol for security monitoring, and inspection of national parks.

Drones for Commercial Aerial Survey Applications

Surveying experts believe that aerial surveying of large areas such as national parks is most practical and economical using drones or small UAV platforms. Surveillance is the critical component of commercial surveying. Surveillance applications include livestock monitoring, wildfire mapping in minimum time, gas and oil pipeline security inspection, home premises security, highway road patrol, and antipiracy missions. Increased development of automatic and sophisticated electro-optical and electromagnetic sensors will allow the drones to conduct high-resolution commercial aerial surveillance missions with minimum cost and complexity.

Drones for Remote Sensing Applications

Drones equipped with appropriate sensors could permit remote sensing functions with minimum cost. Remote sensing sensors include electromagnetic spectrum analyzers, gamma ray sensors, and biological and chemical sensors with minimum size and weight. Electromagnetic sensors include near-infrared and infrared cameras and millimeter side-looking radars with high angular and range resolution capabilities, microwave and ultraviolet spectrum sensors, biological sensors capable of detecting the presence of various microorganisms, and chemical sensors capable of detecting the presence of harmful chemical gases. Laser spectroscopic technology is widely used by chemical analysts to determine the concentration of each chemical agent present in the air.

Drones for Motion Picture and Filmmaking

UAV videography technology is widely used in the United States and European nations. However, the FAA and the European counterparts have not issued formal guidelines regarding drones in the private sector. It should be noted that the explosive growth of crowded filmmaking sites has unquestionably caused headaches for the legislators on both sides, and the bureaucrats are unwilling to handle the issue efficiently and quickly. Note the FAA is debating the requirements for the guidelines, while the European regulators are meeting to iron out the rules and regulations for UAV operations in European aerospace. Domestic lobbyists are petitioning for the use of UAVs or drones for commercial uses including photography, videography, and surveillance functions [6].

Manufacturers of popular commercial UAVs and aerial photography equipment claimed that the FAA regulations normally permit hobbyist drones when they fly below the authorized altitude of 400 ft above the ground and within the UAV operator's line of sight. Film industry experts believe that the use of drones or UAVs for filmmaking is generally easier on large private lands or in rural areas with fewer space concerns for safety or terrorist activities. The FAA stated in 2014 that it has received a petition from the Motion Picture Association of America seeking approval for use of drones in video and filmmaking business activities. Filmmaking experts believe that low-cost drones are best suited for film shots that would otherwise require a helicopter or conventional manned aircraft, which will involve heavy expenses and more time to complete the project. Because of these facts, drones are widely deployed by moviemakers in other parts of the world.

Drones for Sports Events

Besides their applications in photography and cinematography activities, drones have been proposed for the broadcasting of sports activities. Therefore, persons who cannot afford to attend the sports event now can see sports events on their television sets in their homes. Drones have been deployed in the 2014 Winter Olympics in Sochi (Russia) in filmmaking, skiing, and snowboarding events. The principal advantages of using drones or UAVs in important international sports are that they can get closer to the international athletes and they are more flexible than cable-suspended video cameras.

Role of Drones in Domestic Policing Activities

Reliable news broadcasters and law enforcement agencies believe that UAVs or drones are more suitable for domestic policing activities. Drones equipped with high-resolution video cameras can track running criminals or criminals driving high-speed stolen automobiles regardless of locations and tactics. Current TV news shows how fugitive criminals are caught, apprehended, and arrested with minimum time and without firing a shot. A few years back, a Los Angeles Police helicopter tracked O.J. Simpson using helicopter-based video cameras, as Simpson drove recklessly on California freeways. UAVs or drones are widely deployed in the United States and Canada in policing activities because the police of these two countries are well trained in such activities. Lately, dozens of U.S. police forces have applied in 2013 for UAV permits for undertaking such activities. Some Texas politicians and liberal social leaders have warned about potential privacy abuses from aerial surveillance and reconnaissance missions tracking civilians without court approval. Furthermore, Seattle's mayor responded in 2013 to public protests by scrapping the Seattle Police Department plan for the use of drones or UAVs for domestic policing activities [7].

It is interesting to point out that the North Dakota police arrested a criminal engaged in stealing cattle in a rural setting. This arrest was made possible with UAV-based surveillance and reconnaissance data collected using high-resolution video cameras. Again UAV technology has played a critical role in the arrest of the criminal who was sentenced to a 3-year prison term in January 2014. As a matter of fact, this case gained national attention because it was the first time a law enforcement agency used drone technology in carrying out the arrest of the criminal with no collateral damage or physical harm to police officers.

Drones for Oil, Gas, and Mineral Exploration and Production

Surveying authorities believe that UAVs or drones can be used in geomagnetic surveying efforts using the precision movements of the

Earth's differential magnetic field strength, which helps in calculating the nature of the underlying rock structure. A thorough knowledge of the underlying rock structure permits the geophysicists to locate the exact presence of mineral deposits. It is believed that drones can play a critical role in identifying and mining the deposits of rare earth elements, which are widely used in electrical and hybrid electrical vehicle parts, infrared lasers operating at room temperature (uncooled lasers), rechargeable batteries, and a host of electromechanical sensors. Oil and gas companies believe that UAVs can play an important role in the efficient production of oil and gas by monitoring the integrity of oil and installations of gas pipelines. UAVs have been used to monitor the critical activity parameters using the digital cameras mounted under the belly of the vehicle.

UAVs for Disaster Relief Activities

Disaster relief agency management authorities believe that UAVs can provide multiple types of disaster relief capabilities. UAVs can deliver emergency drugs, vaccines, medical equipment, and doctors efficiently and quickly to remote and inaccessible affected areas, where conventional aircraft or trucks are unable to provide medical emergency services. This kind of speedy service can save human lives with minimum cost. Furthermore, UAVs or drones are capable of producing pictures of the real situation for broadcast stations, verifying services provided to the people affected, and giving recommendations for the medical personnel on how to direct their resources to mitigate damages.

Drones for Scientific Research in Atmospheric Environments

Drones or UAVs have demonstrated that atmospheric research scientists will benefit the most in conducting their research activities with minimum cost and risks. Atmospheric research scientists feel that they can handle scientific research activities under any atmospheric environment. The National Oceanic and Atmospheric Administration has deployed Aerosonde UAVs in 2006 as a hurricane hunter tool. Aerosonde Private Ltd of Victoria (Australia) has manufactured the 35-lb UAV, which can fly into the hurricane environment and provide real-time data directly to the National Hurricane Center in Florida. Note the standard barometric pressure and temperature can be obtained from manned hurricane hunters such as Northrop Grumman RQ-4 Global Hawk UAV for hurricane extended measurements. Smart software aboard the UAVs or drones will allow plotting the disaster relief parameters accurately. These data can be made available to national weather reporters and TV news broadcasters such as NBC, CBS, and ABC. In other words, UAVs or drones can offer weather reports in advance before the arrival of a hurricane, thereby minimizing the loss of property as well as human lives. A UK manufacturer produces the variant of their 20-lb Vigilant lightweight drone. This particular UAV or drone is strictly designed for undertaking scientific research missions in severe climatic regions such as Antarctica.

Classic Example of Search and Rescue Mission   A classic example of search and rescue mission conducted by a UAV in remote, inaccessible regions will benefit the most to the readers and to public in general. Deployment of a drone or UAV was demonstrated during the 2008 hurricane, which struck the states of Louisiana and Texas. According to these state authorities, micro-UAVs such as the Aeryon Scout have been deployed to undertake search and rescue operations at moderate altitudes. However, for higher-altitude operations between 20,000 and 30,000 ft, Predator series UAVs will be best suited for performing search, rescue, and damage assessment missions during a 24-hour period, which might involve higher operating costs. Note that these Predators are equipped with electro-optical sensors, high-resolution IR cameras, and mini side-looking radars (SAR), which are best suited to obtain high-resolution images of the targets regardless of weather conditions. The coherent detection capability of a SAR yields exceptionally high-quality images, which are considered most desirable for search, rescue, and damage assessment missions.

UAVs or Drones for Animal Conservation Functions

Countries like Nepal, Tanzania, and South Africa are using drones for animal conservation activities [8]. Coal, gas, and mineral authorities are contemplating such activities in Vietnam, Tanzania, and Malaysia. An international welfare organization recommended training some personnel to use drones for antipoaching operations. In 2012 UAVs were deployed by the Sea Shepherd Conservation Society in Namibia to monitor and document the annual seal slaughtering operation. In December 2013, a Falcon unmanned aerial vehicle system was selected by the Namibian government to help in combating rhino poaching operations. The drones equipped with exotic technology devices, such as a radio-frequency identification device, have been recommended to monitor the rhino population in Etosha National Park, Namibia.

Drones for Maritime Patrol Activities

Recently some countries, particularly People's Republic of China, is making a bogus ownership claim over their neighbor's properties, which are historically, ethnically, and geographically significantly different from China. Various countries in South Asia have expressed serious concern and anger at this Chinese behavior. Classic examples of Chinese claims include the Indian State of Arunachal Pradesh; isolated islands very close to Japanese coastal regions, which are very far from the Chinese coast; Vietnam islands rich in gas, oil, and mineral deposits; and islands extremely close to the Philippine coast. Countries like Japan, South Korea, Vietnam, India, the Philippines, and Brunei are in territorial disputes with China. These countries are in serious legal disagreement with China due to its aggressive, hegemonic, and assertive attitude. According to international press reports, countries such as Vietnam and Japan are willing to go to war with China on this particular subject, which could disturb international peace and security. China has illegally as well as militarily invaded and occupied Tibet since 1959 when Mao killed millions of innocent Tibetans. International papers have expressed shock and called Mao the butcher of Lassa, and Zhou Enlai the most notorious leader of China, who was the principal architect of the 1962 Indo-China War. Note there is great tension between the Chinese government and the governments of Southeast Asian countries over oil, gas, and coal deposits claimed by the Chinese, which are near the Vietnam coast. Note that the Chinese government exclusively claims rights on oil, gas, and coal deposits in the South China Sea, which is in contrast with the UN Charter that clearly declares that all countries of the world have unrestricted navigation rights in the South China Sea without any hindrance.

Drones for Cooperative Forest Fire Surveillance Missions

A close examination of published literatures indicates that recent fires have caused billions of dollars of damage to property and environment every year. Environmental scientists believe that to combat forest fires effectively, their early detection and continuous tracking are of critical importance. The scientists further believe that with the help of advanced image processing techniques, several methods have been developed to detect forest fires in remote regions using satellite images and high-resolution images collected by the electro-optical sensors, compact SARs, and wideband IR detectors aboard the UAVs. Effective detection of forest fires is possible using a three-dimensional histogram described in the *2002 IEEE International Geoscience and Remote Sensing Symposium.*

Such images can also be captured by low-Earth orbiting satellites, an orbital period of 10 hours approximately, and with a resolution that is just sufficient for detection. Firefighters need frequent and high-quality data updates on the progress of fires for effective control of the fires. In case the forest fire monitoring techniques are deficient, firefighters are often required to enter the fire regions with little knowledge of how and where the fire is propagating, which could put their lives at risk. Firefighting experts believe that UAVs or drones equipped with the latest electro-optical sensors and high-resolution IR cameras are best suited for such activities. In summary, the speed of fire detection resolution of electro-optic sensors and high-resolution IR detectors are the most important parameters. To make the platforms for surveillance more effective and reliable, there is an urgent need to develop more effective fire monitoring techniques.

NASA Contribution to Firefighting Technology   Surveillance scientists feel that HALE UAVs such as the Altair have more potential for enhanced range resolution and higher update rates over the satellite-based systems. However, the limited availability of HALE systems during the peak fire seasons may limit their overall effectiveness and thus emphasizes the need for low-cost locally available surveillance systems. The low-altitude, short-endurance (LASE) UAVs are expected to be a key technology to improve fire monitoring applications. Note that when flying at low altitude, these UAVs can capture high-resolution imagery and forecast frequently to commercial weather broadcasting stations such as NBC, ABC, and CBC for the general public interest.

Weather experts believe that pursuing this possibility with ongoing research projects aimed for attacking the growth of fires using LASE-based UAVs will permit the weather forecasters to provide more accurate weather reports. NASA research scientists are interested in demonstrating "UAV-Based Over-the-Horizon Disaster Management Demonstration Capability."

NASA scientists believe that the LASE UAV is low-cost technology that enhances fire monitoring. However, a number of challenges have to be solved before LASE-based UAVs can be deployed for effective fire monitoring operations. First, with fire growing in intensity and changing directions, UAVs must be capable of planning their path using limited near-time information. Second, LASE-based UAVs typically cannot carry fuel to endure a long firefighting mission, which means that the UAV must have the intelligence to return to home base for refueling. For large forest fires, the information update rate may still be too low even if a single LASE unmanned aerial vehicle is deployed for a fire monitoring mission. One has to face the unavoidable situation when the fire generates a tremendous amount of heat and turbulence directly over the burning region. Under such conditions, crossing directly over a fire will place the LASE vehicle at significant risk. Such UAVs are effectively restricted to the aerospace over the unburned region of the fire. NASA scientists are exploring the feasibility of using multiple LASE-based UAVs to cooperatively monitor and track the large propagation of large forest fires. Note that by using multiple UAVs for cooperatively monitoring and tracking the propagation of fire in terms of intensity and direction, the complexity of the system will shift from the hardware platform to the cooperative control strategies used to coordinate fire monitoring operation. While teams of LASE UAVs will be more robust to single point of failure than a single satellite or the HALE UAV, several technical challenges must be addressed to enable their successful implementation. Issues addressed so far include overcoming limited communication range and flight duration, developing a suitable coordination strategy for fire monitoring, and forming team consensus in the presence of noisy or incomplete information. A recent increase in interest in UAV activities sponsored by the military community has started funding intensive research studies in the field of cooperative control of UAVs or drones. There are advanced research studies in the field of cooperative search of multiple UAVs under collision avoidance and limited range communication constraints. These research activities will address real-time search and task allocation in the UAV teams and cooperative path planning for multiple UAVs in dynamic and uncertain operating conditions. According to UAV design engineers, cooperative control of multiple UAV vehicles will become not only more complicated but also unmanageable under severe dynamic and poor communication environments.

Current research activities have been focused on cooperative real-time search and track allocation for UAVs operating in team formation, Over-the-Horizon Disaster Management Projects under dynamic and uncertain flight conditions, formation flight test results for a UAV research aircrafts, and decentralized optimization with applications to multiple air platform coordination with particular emphasis on "multiple-vehicle rendezvous problems. Advanced research activities must focus on experimental demonstration on semiautonomous control technology dealing with team consisting of multiple UAVs and coordination and control experiments on a multivehicle test bed.

Cooperative Forest Fire Surveillance Using a Team of Micro-UAVs

A thorough understanding of the cooperative forest fire surveillance technology using a team of small UAVs or micro-UAVs requires a comprehensive examination of published articles and models. Most of the information is available from the technical articles in the *International Journal of Systems Science* (volume 37, issue 6, 2006, particularly, Ref. [9]).

Benefits of cooperative forest fire surveillance technology using a team of small UAVs or micro-UAVs will be discussed in great detail but in a logical fashion. The following steps must be taken into consideration:

- Problem statement
- Fire perimeter tracking for a single UAV platform
- Cooperative team tracking
- Simulation of data collection and evaluation
- Base station capability to deploy multiple UAVs to monitor the propagation of the fire in terms of intensity and wind direction
- Conclusion

The principal objective of this particular study is to explore the feasibility of deployment of multiple LASE UAVs and to cooperatively monitor and track the propagation of large forest fires. A real-time algorithm is best suited for tracking the parameters including intensity and direction of the fire, which can be monitored by the onboard IR sensor.

Real-Time Algorithm   Using this algorithm, a decentralized multiple UAV approach can be developed to monitor the parameters of the fire. These UAVs are assumed to have limited communication range and sensing capability [9]. The effectiveness of the technique can be demonstrated by a simulation involving a six-degree-of-freedom dynamic model for the unattended aerial vehicle and a numerical propagation model for the forest fire. Note that a six-degree-of-freedom dynamic model is essential to demonstrate the effectiveness of the model. Note that the important features of this technique or approach include the ability to monitor the changes in the fire parameters, the ability to systematically add or remove the UAVs from the original team, and the ability to supply time-critical information or fire parameters to the firefighters. As mentioned earlier, about overall effectiveness and low cost, LASE UAV technology will be found most suitable for enhanced fire monitoring capability.

Evidence of Experimental Demonstration of UAV Team   Experimental work involving teams of UAVs has been very limited because of practical challenges of fielding multiple UAVs simultaneously, operational costs, and logistical problems. Several research scientists have demonstrated the leader following with two small UAVs. Such research activity provided the experimental demonstration of semiautonomous control of UAVs. During the research studies, some problems were observed. Agreement problems have been mentioned in networks with directed flight paths and switching technology. Coordination variables and consensus building in multiple vehicle systems have been observed in the experimental flight tests. Some technical published papers have discussed the benefits of multiple LASE UAV cooperative control solution techniques for forest fire monitoring problem. One particular published document suggested the development of a real-time algorithm for precise tracking of the perimeter of fire when an onboard infrared sensor is available. A decentralized multiple-UAV technique to monitor the perimeter of the fire is recommended by various research scientists. The salient features of this approach can be summarized as follows:

- The ability to monitor an instantly changing fire perimeter
- The ability to systematically add and remove UAVs from the fire team, which is considered critical for refueling of the team UAVs
- The ability to supply the time-critical information to the firefighter

Experimental results observed during the simulation can be briefly stated. A realistic simulation of the real evolution of a typical forest fire has been demonstrated using the forest fire propagation model in 1996 by the research scientists Gardner and Hargrove. This particular model has been used by various scientists in the development of cooperation algorithm verification of path planning and simulation of forest fire scenarios. Forest fire scenarios can be defined involving certain assumptions, which are as follows:

- It is assumed that each UAV can collect or receive sufficient information on board to plan and adjust its flight path autonomously. This assumption will allow the UAV to adapt its aerodynamic parameters that are carefully selected and optimized by the most experienced and seasoned aerospace pilots. In summary, the salient features of this particular approach for a typical forest fire monitoring scenario can be analyzed as follows (Figure 1.4):

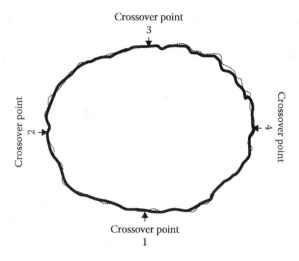

Figure 1.4   Fire monitoring scenario using multiple UAVs to monitor the propagation of fire at the base. The curve represents the fire parameters at four crosspoints. (Adapted from Casbeer, D.W. et al., *International Journal of Systems and Science*, 37(6), 351–360, 2006.)

- The number of UAVs deployed should not exceed three.
- All UAVs must be well informed about formation flight.
- Each UAV must have high-performance infrared cameras installed on the front, rear, bottom, and along the fuselage sections with clear views of the targets at all times. These IR cameras must collect images of a small region below the UAV. The IR camera is particularly designed for fire monitoring purposes as it detects the regions of the ground with highest temperatures.
- For optimum fire monitoring capability, each UAV in the team must know the flight pattern and the fire parameters to coordinate with each other for optimum firefighting efficiency.
- Each UAV is assumed to have a limited communication range, which means that it cannot upload data to the base station unless it is within a certain range from the base station, and it also cannot communicate with other UAVs operating within the proximity of the base.
- Finally, each UAV in the team is assumed to have limited fuel capacity, which specifically implies that the UAV must periodically return to the base station for refueling purposes.

- One can expect delay between the images that are collected and transmitted by the team UAVs. Time delays between the images are collected and, when they are transmitted to the base station, can serve as a matter of the quality of the fire monitoring algorithm. As a matter of fact, the quality of the fire monitoring algorithm is recognized as the performance parameter of the firefighting approach or scheme.

Development of a Cooperative Surveillance Strategy   Development of a proven cooperative surveillance strategy plays a critical role in minimizing the latency associated with fire perimeter measurements delivered to the base station, which keeps the track of fire perimeter measurements contributed independently by each firefighting vehicle. As mentioned earlier, the delay between the images collected and transmitted to the base station can serve as a measure of quality control parameter for the cooperative surveillance capability or as a measure of fire monitoring algorithm.

Assume $\Delta(x, t)$ the complex latency parameter associated with the information about the position x along the perimeter at time t. As time passes, the information at the base station grows or becomes more latent until another UAV arrives to transmit the latest information it has gathered using its IR camera. Note for a particular position x along the fire perimeter, the complex latency parameter will increase with time until it is replaced by the data just downloaded by another UAV. The latest downloading of data by another UAV gives a typical depiction of latency evolution for a particular point x_0 on the perimeter of the fire as shown in Figure 1.5. The vertical edges of the sawtooth waveform represent the transmission of the data provided by UAV to the base station. Note that the linearly increasing portion of the waveform represents the increase in the latency of UAV updates. The minimum latency (Δ_{min}) corresponds to the time of flight between the point of interest and the base station. It is further interesting to know that the maximum latency is strictly dependent on the total time required to make an observation at the distance x_0 and to deliver the data to the base station. Note that the development of cooperative surveillance technology

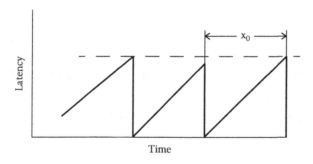

Figure 1.5 Latency between updates for a point x_0 on the perimeter of the fire. (Adapted from Casbeer, D.W. et al., *International Journal of Systems and Science*, 37(6), 351–360, 2006.)

is essential to minimize the latency associated with fire perimeter measurements delivered to the base station by the team vehicles. This is accomplished by minimizing the flight time between the points shown on the fire perimeter and the base station and by maximizing the frequency of measurement updates delivered to the base station. We will now focus on critical steps, namely, fire perimeter tracking for a single firefighting UAV, cooperative firefighting team tracking, latency minimization, and a distributing fire monitoring algorithm.

Critical Aspects of Fire Monitoring Scheme Based on Autonomous Concept So far, critical aspects of the firefighting scheme have been described and evaluated involving a team of UAVs, each being controlled by the ground control operator. It is possible for the ground control operator to control the performance of a specific UAV and to response to another UAV of the firefighting team. Studies performed on autonomous technology indicate that it offers significant benefits on latency minimization and virtually error-free fire monitoring algorithm, which is not possible using the previous operating technique.

It is critically important to summarize the benefits for the fire perimeter tracking using a team of few unmanned aerial vehicles and the autonomous technology [9]. Cooperative fire monitoring strictly relies on the ability of each individual UAV to track the fire perimeter independently. Comprehensive studies performed on the autonomous technology seem to reveal that a robust fire tracking algorithm must be implemented in the computer simulation. The computer simulation must assume that each UAV is equipped with a highly sensitive infrared camera on a pan and tilt gimbal and an autopilot system that is functionally similar to the one described for autonomous vehicle technologies applicable to small fixed-wing UAVs. Technical approach for the firefighting scheme using autonomous technology can be briefly summarized in six discrete steps that are each performed at the frame rate of the onboard infrared camera:

1. Scan carefully through the infrared spectral range, leveling each pixel as (a) burned and (b) unburned.
2. Deploy a linear classifier to segment the image into two portions, with burned elements in one portion and the unburned elements in the other portion.
3. Draw or project the segmentation line room of the camera frame into the geometrical coordinates of the unmanned aerial vehicle.
4. Construct the set of reachable points T second apart into the future; parameterize this set as a function of the roll angle. Specific details regarding this construction have been systematically reported in Beard, R., et al., *Journal of Aerospace Computing, Information and Communication* 2, 92–108, 2005.
5. Give command on the roll angle that corresponds to the point in reachable set that is located at a distance "d" on the unburned side of the fire perimeter.
6. Give a command to pan and tilt angles of the infrared camera gimbal so that the segmentation line of the linear classifier divides the IR image into two equal parts.

Summary of This Approach   UAV experts consider this particular approach as best suited for tracking the perimeter of a forest fire or a high-intensity fire, because it usually evolves in a noncontiguous fashion. Under this particular circumstance, the fires will jump over roads and streams and subsequently spread around rocky terrain. Because fire spreads faster uphill than downhill, according to firefighters, it leads to fingering phenomena in mountain environments, where fires spread in finger patterns along ridges in the terrain. The linear

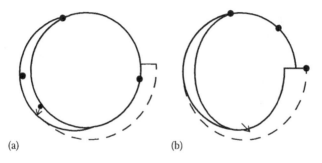

(a)   (b)

Figure 1.6   (a) Latency profile for a single unmanned aerial vehicle (UAV) monitoring a static circular fire. (b) Latency profile for a pair of UAVs monitoring a static fire in opposite direction. (Adapted from Casbeer, D.W. et al., *International Journal of Systems and Science*, 37(6), 351–360, 2006.)

classifier deployed in the earlier paragraph effectively smoothes out through noncontiguous boundaries. The distance "d" ensures that the unmanned aerial vehicle does not fly directly over burning fire, but rather flies at a safe offset distance "d" as illustrated in Figure 1.6. Note that maneuvering the gimbal ensures the continued effectiveness of the linear classifier. In brief, the fire tracking scheme described herein has been successfully demonstrated in computer simulations undertaken by various UAV designers. Computer simulation results will be described briefly in the following section.

Discussion of Computer Simulation Results Pertaining to Fire Model The presentation of computer simulation would essentially highlight the effectiveness of the algorithms developed by various scientists. Computer simulation results under discussion here involving autonomous system elements were obtained by David W. Casbeer and associates [9] with particular emphasis on fire perimeter tracking for a single UAV and cooperative team tracking (Figure 1.7). Simulation results obtained by these scientists highlight the effectiveness of the algorithms developed using the Ecological Model for Burning in the Yellowstone Region (EMBYR), which divides this particular region into a grid of cells, each having specific properties that will affect the spread of the fire. These properties identify the type of foliage, the moisture level in the foliage, and the elevation of the foliage. At a given time step, the fire will spread from a burning cell to nonburning cells according to an independent stochastic event that is a function of the properties of the respective cells. By running the EMBYR

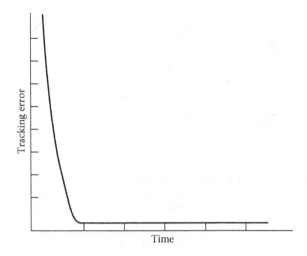

Figure 1.7 Fire perimeter tracking error for a single unmanned aerial vehicle, which was commanded to track the fire with an offset of 100 meters. (Adapted from Casbeer, D.W. et al., *International Journal of Systems and Science,* 37(6), 351–360, 2006.)

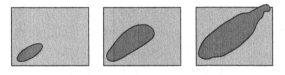

Figure 1.8 Output of the Ecological Model for Burning in the Yellowstone Region simulation program showing the fire simulation under high wind conditions. (Adapted from Casbeer, D.W. et al., *International Journal of Systems and Science,* 37(6), 351–360, 2006.)

program multiple times and averaging the numerical results, one can achieve realistic fire simulations like the one shown in Figure 1.8. The fire simulation under high wind conditions as a function of elevation gradient is evident from Figure 1.8. It is clear from this figure how the fire is spreading in the direction of the wind. Note that the earlier discussion is limited to the fire model only.

Assumptions for Fire Monitoring Technique   Some assumptions are needed to monitor the fire perimeter. The following assumptions are necessary for effective fire monitoring:

• It is assumed that each UAV can collect or receive sufficient information on board the vehicle to plan and adjust the UAV flight path autonomously. This allows the UAV to adapt its flight path according to the fire perimeter. Essentially, each UAV is assumed to be equipped with an IR capable of capturing images beneath its path. An infrared camera is best suited for fire monitoring as it detects the regions of the ground with highest temperatures.

- Second, each UAV taking an active part is assumed to have a limited communication range, which means that it cannot upload data to the base station unless it is within a certain operating range of the base station, and it cannot communicate with other UAVs unless it is within a specified range.
- Finally, each UAV is assumed to have limited fuel, which implies that it must periodically return to the base station for refueling.

Note that the delay between when the images are collected and when they are transmitted to the base station can serve as a measure of the quality of the fire monitoring algorithm. Let $\Delta(x, t)$ represent the latency associated with the information about the position x along the perimeter at a time t. As time passes, the information at the base station grows older or becomes more latent until a new UAV arrives to transmit the latest information it has gathered as illustrated by Figure 1.5. For a particular position x along the fire perimeter, the parameter $\Delta(x, t)$ will simply increase with time until it is replaced by the data downloaded from another UAV. Figure 1.7 gives a typical depiction of latency evolution for a particular point x_0 on the perimeter of the fire. The vertical edges of the sawtooth waveform shown in Figure 1.9 indicate the transmission of data from the UAV to the base station, while the linearly increasing portion of the waveform represents the increase in latency between the UAV updates. Note that the minimum latency Δ_{min} corresponds to the time of flight between the point of interest and the base station. The maximum latency depends on the total time required to make an observation at the location x_0 and to deliver that data to the base station. This particular simulation effort employs a cooperative strategy that minimizes the latency associated with the fire perimeter measurements provided to the base station. This is accomplished by minimizing the time of flight between the points of the fire perimeter and the base station and maximizing the frequency of measurement updates provided to the base station.

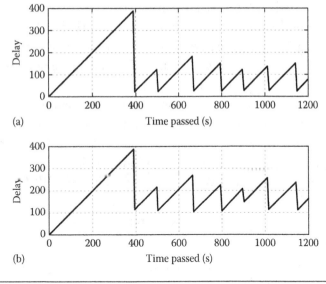

Figure 1.9   Vertical edges of the sawtooth waveform represents the increasing latency between the unmanned aerial vehicle updates. (a) UAV-1 update, (b) UAV-2 update. (From Casbeer, D.W. et al., *International Journal of Systems and Science*, 37(6), 351–360, 2006.)

Minimization Technique for Latency   Minimization of latency is essential for effective fire monitoring irrespective of UAVs operating in the team. When a UAV transmits its information to the base station, an associated latency profile accompanies the data.

Let the latency associated with a point x of the base station update the information at the time of the base station and should be denoted by function p(x). It is important to note the function $\Delta(x, t) = p(x)$ when a UAV updates the information at the base station. In other words,

$$\Delta(x, t) = p(x) \quad \text{when the UAV updates the information} \quad (1.1)$$

and

$$\Delta(x, t) = [p(x) + (t - t_{update})] \quad \text{between updates} \quad (1.2)$$

where p(x) is the latency profile.

Studies performed on cooperative monitoring scheme indicate that the principal objective for the data updating is to design the cooperating monitoring technique, which minimizes the function p(x) for every x and updates the base station as often as possible.

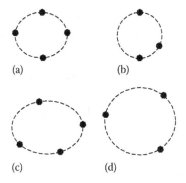

(a)   (b)

(c)   (d)

Figure 1.10 Latency associated with four UAVs engaged on the perimeter of the fire. (a–d) Latencies for each UAV.

Figure 1.10 shows the latency associated with the perimeter of the fire when a clockwise flying UAV arrives back at the base station after the complete tour of the perimeter, where the thickness of the path indicates the latency of the base station update of that point. The state of the fire is transmitted only after the UAV traversed the entire fire perimeter; the optimum latency is associated with the data gathered at the beginning of the flight near the base station. Since the UAV is traveling at a constant velocity, the latency profile is a linear function of the distance travelled p(x), which is defined by the following equation:

$$p(x) = \frac{P - x}{v} \tag{1.3}$$

where
P is the perimeter of the fire
v is the velocity of the UAV

Note the base station receives updates only as fast as the UAV can traverse the entire fire perimeter. Therefore, the total latency associated with one traverse can be given by the following equation:

$$\text{Total latency} = \int_0^P p(x)\,dx = \frac{0.5\,P^2}{v} \tag{1.4}$$

Latency profile for a single UAV monitoring the static circular fire is illustrated in Figure 1.6. Note that the thickness of the path indicates the latency information at the point when it is transmitted to the base station. Furthermore, the latency profile for a pair of UAVs monitoring a circular static fire in opposite directions can be seen in Figure 1.6.

For better understanding of overall latency, let one UAB be assigned to survey the upper half of the perimeter while the second UAB assigned to lower half. If the UAVs depart from the base station simultaneously and fly at the same speed, the update rate will be same as the single UAV case (assuming that both UAVs arrive back at the same base station at the same time), but the latency associated with the information on both sides of the base station will be symmetric and reduced as illustrated in Figure 1.4. In this case, the latency profile can be expressed by the following equation:

$$p(x) = \frac{x}{v} \quad \text{for } 0 < x < \frac{P}{2} \tag{1.5}$$

$$p(x) = \frac{p - x}{v} \tag{1.6}$$

But the overall latency associated with this scheme is as follows:

$$\text{Overall latency} = \int_0^P p(x)\, dx = \frac{0.25\, P^2}{v},$$

$$\text{which is half that of the single UAV case} \tag{1.7}$$

For UAVs that follow the perimeter and fly at constant velocity (v), the latency profile shown in Figure 1.5, which is the minimum possible latency associated with data gathered at point x on the perimeter, is the same needed to travel from x to the base station along the shortest path. Dividing the perimeter equally between two UAVs ensures that the distance traveled between any point on the perimeter and the base station is minimum. The result is that adding more than two UAVs will not improve the latency profile p(x) for any particular point on the perimeter. Furthermore, the rate at which the update occurs will increase linearly with the number of UAV pairs deployed, and consequently, the effective latency will be reduced. It is further important to mention that it is necessary to maintain the minimum latency profile as shown in Figure 1.5 and to maximize the frequency of updates at the base station. For further latency improvement, the UAVs should be equally distributed around the fire perimeter with each UAV assigned in order to monitor a segment of length P/N, where P is the perimeter of the fire and N is the number of UAVs involved.

Simulation Results Pertaining to Perimeter Tracking   The previous step described in the fire tracking algorithm scans through an image from an infrared camera and to identify each pixel as burned or unburned. In this particular simulation, the EMBYR fire model is used to generate an a priori model of the fire at each time step. During the execution of the simulation, the current simulation time and the state of UAV are used to generate a binary image where 0 represents an unburned element and "1" represents a burned element. Simulation results of a single UAV tracking a fire perimeter are shown in Figure 1.7. The commanded offset or safe distance from the boundary of the fire was 300 ft or 100 m. From Figure 1.7, one can see the fire tracking error for a single UAV. In this particular computer simulation, the UAV was asked to track with an offset distance of 100 m to maintain utmost safety in actual practice. As a matter of fact, the fire perimeter will not be contiguous due to noncombustible regions such as lakes or boulder fields. However, if the field of view (FOV) of the IR camera is larger than the gaps in the fire perimeter, then the algorithm will function accurately because the linear classifier will provide the best linear fit to the simulation data, which will fit a line through the noncontiguous regions in the fire perimeter.

Simulation Data for Cooperative Tracking   Under this particular heading, the simulation results or data will be provided involving multiple UAVs to monitor the perimeter of a forest simulated using the EMBYR model. Note that the maximum communication range for the UAVs was selected 100 m, which is approximately 9 pixels.

The velocity of the UAVs was set at 18 m/s. The fire perimeter was growing at an average of 2.8 m/s. Initially, four UAVs were assigned to monitor the fire, and later on two more UAVs were added as shown at the bottom left in Figure 1.10, approaching the fire from the base station. A short time later, as illustrated in Figure 1.10b, the two new UABs were loitering while waiting for their next rendezvous. After approximately 10 min has passed, the UAVs are in steady-state configuration as shown in regions (c) and (d) in Figure 1.10.

In the fire monitoring simulation, the UAV configuration does not precisely converge in the equilibrium predicted by the algorithm due to the dynamic characteristics of the UAVs as well as of the fire. Remember that the steady-state configuration error depends on the turning radius of the UAVs and the growth rate of the fire perimeter.

Despite these problems, the lengths $L_i(k)$ converge as time progresses. It can be seen in the region of Figure 1.10 that these lengths are increasing as time progresses due to the growth of the fire. Regions (b) and (c) of Figure 1.10 show the convergence of $L_i(k)$ static fire with perimeter of length 7.2 km. In the region (b) of Figure 1.10, four UAVs initially monitoring the fire are shown, and after 1000 s, two more UAVs were introduced. In conclusion, there are six UAVs involved in monitoring the fire in region (c) of Figure 1.10.

Potential Algorithms for Fire Monitoring Purposes   The studies performed on various algorithms indicate that they are best suited for distributing fire monitoring schemes. The studies were limited to two or three algorithms capable of providing minimum latency for fire monitoring schemes with minimum error. For evaluating the effectiveness and suitability algorithms, a distributing fire monitoring scheme was selected. For a fixed perimeter length and a fixed number of UAVs deployed, the minimum latency configuration occurs when pairs of UAVs are uniformly spread along the perimeter of the fire in both directions (i.e., for every pair of UAVs, one UAV is headed for clockwise direction and the other UAV for counterclockwise direction). Note that these pairs of UAVs will meet and transmit the gathered information, and then UAV will reverse its directions to meet its neighbor in the other direction. In order to facilitate refueling, the UAVs can exchange roles at a rendezvous so that the team members with least fuel are nearest to the base station.

Distributed Algorithm   The principal objective of the distributed algorithm is to converse to minimum latency configuration with minimum computational efforts. This algorithm must converse for any fire perimeter size and must readjust when the perimeter length or the number of UAV pairs varies. It is possible to design the algorithm so that the changes in the system parameters are propagated across the UAV team as rapid as possible. This will permit the algorithm designer to address the insertion and deletion of UAVs in the team. In addition, this algorithm will allow the expansion and contraction of the fire perimeter.

The fundamental requirement is for each UAV to take appropriate action that will allow the neighboring UAV pairs to share the perimeter between them, improving the latency. When two UAVs meet, each has clear knowledge of the length just traveled from its previous rendezvous position. Note that the sum of these lengths can be divided equally between the UAVs so that the UAV that has traveled the least distance will loiter at the midpoint of this segment to wait for its neighbor the next time when the two UAVs are supposed to meet.

To illustrate this concept, let us consider a simple line segment of length L with two UAVs, which are tasked for gathering information cooperatively along the line segment. Let L_i be the distance traveled by the ith UAV from the last endpoint it has visited. Further, assume that the first UAV has traveled to a least distance, so after returning to the endpoint, it will travel to a distance equal to $(L_1 + L_2)/2$ and then begin to loiter. In the meantime, UAV_2 will return to endpoint and then reverse its direction until it meets or encounters UAV_1. Since UAV_1 has traveled the shortest distance, it will arrive at the midpoint first, and when both arrive, each UAV has traveled the same distance (which means $L_1 = L_2$) from the endpoints, which means that the UAV pair has achieved minimum latency configuration. This means this example demonstrates the performance of two UAV pairs monitoring a fixed fire perimeter tracking performance. Note that any change in the size of the segment will be tracked because the UAV pair essentially measured the current perimeter length by summing the distance traveled from the endpoints. In other words, the UAVs only have memory of the state of perimeter from one previous iteration; a continuous load balancing algorithm will track finite changes in the fire perimeter.

Load Balancing Algorithm   The load balancing algorithm plays an important role in balancing the travel distance loads along the fire perimeter. By measuring the discrepancy in segment distance L_1, and the new distance back to the endpoint, the UAV_1 can update the loiter distance to neutralize the negative effect of the growth in that region. Note that adding UAVs to the perimeter is tantamount to stringing the perimeter segments together. This will have changing endpoints. The endpoints shared by one pair of UAVs are the outside neighbors of these UAVs. Now one needs to deploy Monte Carlo simulation to verify that pair-wise load balancing will lead to team convergence completely. One must maintain that by balancing the length shared by every pair of UAVs, the team as a whole will spread itself evenly around the fire perimeter in order to achieve the minimum latency configuration. If the algorithm can be shown to converge for artificial initial conditions with an arbitrary number of team members, then the insertion/deletion can be analyzed by considering the modified system (after the insertion and deletion processes) with initial conditions given from the state of original system at the time of insertion/deletion. Note that each UAV must implement the following condition or requirements to the load balancing algorithm:

- Maintain a balanced estimate of the distance from the last rendezvous in each direction. Note that each UAV shares a segment with its clockwise neighbor and its counterclockwise neighbor.
- At a rendezvous position, the UAV that has traveled the smallest distance since its last rendezvous position agrees to loiter at the midpoint of the shared segment the next time it is tracking the perimeter in that direction (either clockwise or counterclockwise direction).
- If the endpoint of the segment is changed due to fire perimeter growth or neighbor actions, then the loiter distance must be augmented by the change in distance of the endpoint. This process keeps the loiter point at the same position relative to the segment length as communicated during the rendezvous period.

- The algorithm operating condition requires at least one UAV in rendezvous pair due to the larger travel distance; the UAV must not loiter en route to the next anticipated rendezvous of this pair. This requirement ensures that the pairs of UAVs will always meet again, irrespective of the change in the fire perimeter.

It can be demonstrated through the Monte Carlo simulation that the load balancing algorithm converses to the minimum latency configuration for arbitrary initial conditions. Consider a simulation instance consisting of launching of N pairs of UAVs from the base station at random times around a fixed length of circular perimeter. It is important that each member of the team continuously balances the load shared with each of the two neighbors. This simulation continues until all UAVs are with the mean standard deviation (ε) of the minimum agency configuration or until the maximum time is reached.

The analysis requires that for each N, the total number of UAVs in the team, the mean standard deviation ε ranging from 2, 3, 4, 5, 6, 7, to 100,000 simulations, was performed and the time required for each steady-state simulation is recorded. Since time to convergence is a function of speed of UAVs and the size of the fire perimeter, the convergence time is normalized by the time required for the information to travel around the fire perimeter. For example, if the convergence time is assumed as T, then one UAV could transverse the entire fire perimeter T times in the amount of time required for the team to converge within ε to meet the minimum latency configuration. The mean and standard deviations in normalize convergence time over the 100,000 iterations for each N with normalized $\varepsilon = 0.0003$ or 0.03% can be seen in Figure 1.11.

Monte Carlo Simulation Method   Studies performed on the subject concerned reveal that the Monte Carlo method plays a critical role in solving statistical problems by reading in the order of the random digits expressed in 10 equiprobable eventualities. Now taking pairs of digits simulates 100 equiprobable eventualities. An event with probability of 63% may be simulated by the reading of successive pairs, considering as a "success" any pair from 00 to 62. The successive pairs

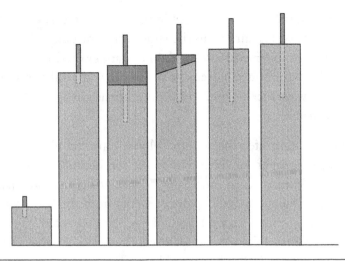

Figure 1.11 Convergence of load balancing algorithm using Monte Carlo simulation involving multiple unmanned aerial vehicles. (Adapted from Casbeer, D.W. et al., *International Journal of Systems and Science*, 37(6), 351–360, 2006.)

divided by 100 approximate a random variable uniformly distributed over the interval (0, 1). But for a smooth approximation, three or four consecutive digits could be used.

Any continuous variable x with cumulative distribution function P(x) can be transformed into a new variable, which is uniformly distributed between 0 and 1 by means function defined by the following equation:

$$r = [P(x)] \qquad (1.8)$$

Conversely, this variable can be simulated by solving the following equation:

$$\text{Distribution function, } P(x) = [r_i], \qquad (1.9)$$

for x where the r_i are successive random numbers

Conversely, the parameter x can be simulated by solving Equation 1.9. Note that the quantity r_i is the successive random number.

To illustrate the application, select 0.49, 0.31, 0.97, 0.45, 0.80, etc. The corresponding values of x, which will be normally distributed with mean zero and variance one, are as follows:

$$0.0, -0.5, 1.9, -0.1, 0.8.$$

This example simulates the results of successive shots aimed at the point $x = 0$. According to simulation scientists, to achieve accurate numerical results, a large number of trials N will be necessary because the simulation accuracy increases roughly as the square root of trials or N. Note that where a crude evaluation is required, it can be accomplished with few trials.

Fire EMBYR Model Using Cooperative Control Algorithm In addition, fire tracking model and perimeter tracking model are available to estimate the trends in fire tracking and perimeter tracking. Fire tracking experts have developed the EMBYR model for an effective prediction of the perimeter tracking cooperative control and time-varying fire simulations [9]. As mentioned earlier, the EMBYR model essentially divides the region of interest into a grid of cells, each with specific properties that affect the spread of the fire. Such properties include the type of foliage, the moisture contents, and the elevation. At a given time step, the fire will spread from a burning cell to nonburning cells according to an independent stochastic event, which is a function of the properties of the respective cells. By running the EMBYR model multiple times and averaging the result, one can achieve the realistic fire simulations like the one shown in Figure 1.8. This figure illustrates the fire simulation under high wind conditions with an elevation gradient. It is interesting to observe that the fire is spreading in the direction of the wind.

Conclusions on Forest Fire Surveillance Concept Forest fire surveillance has been introduced as a cooperative and control problem for UAVs, which can communicate only when they are in close proximity with each other. An approach has been presented for fire surveillance using a single UAV equipped with an infrared camera. A cooperative surveillance scheme is discussed, which utilizes an even number of UAVs to minimize the information latency and the frequency of update requirement. The algorithm has been verified using Monte Carlo simulations. Results using a high-fidelity fire model and 6-UAV model have been presented to demonstrate the effectiveness of this approach.

Summary

Historical aspects of UAVs are summarized with emphasis on potential applications, unique design features, and their emergency deployments in remote and inaccessible regions. UAV classifications for various missions and applications are identified with emphasis on reliability, safety, and ease of operation. UAV design and sensor requirements for border patrol, urgent delivery of commercial and military goods, delivery of emergency medical supplies, fire surveillance missions, and a host of other emergency operations. Performance and reliability requirements must be defined for UCAVs with particular emphasis on combat effectiveness, safety, types of miniaturized weapon systems with high kill probability, and IRS mission capabilities. Classifications of miniaturized UAVs are based on their mission objectives with emphasis on mission requirements and their stealth configurations. There are FAA designations and legal regulations for UAV vehicles. Critical design requirements for UAVs capable of providing the low-altitude and high-altitude long-endurance missions must be. According to the article published in Ref [2], military experts see bright future for the hunter–killer UCAVs in future battlefield conflicts.

Performance capabilities, design configurations, and propulsion system requirements for commercial drones are briefly described with emphasis on cost, safety, and reliability. Critical roles of drones for oil, gas, and mineral exploration, cooperative surveillance using multiple UAVs, surveillance missions for maritime activities, and tracking of wild animals are identified with emphasis on cost, reliability, and ease of operation. Critical design requirements of IR cameras are identified with major emphasis on fire image quality at high temperatures. Development of a cooperative surveillance strategy is discussed in great detail with particular emphasis on minimum latency. Computer modeling techniques on fire tracking surveillance are briefly summarized. Computer simulation results are discussed in great detail. Advantages of potential algorithms are briefly discussed with emphasis on minimum latency. Distribution algorithms, load balanced algorithms, and cost-effective fire tracking algorithms used by various sources are briefly evaluated with emphasis on performance and minimum latency.

References

1. Wikimedia, Unmanned air vehicles and their applications, Wikimedia Foundation, Inc.
2. J.R. Wilson, Hunter-killer UAVs to swarm battlefields, *Military and Aerospace Electronics*, 2–6, July 2007.
3. J. McHale, Ground control stations for unmanned aerial vehicles (UAVs) are becoming networking-hub cockpits on the ground for U.S. unmanned forces. *Military and Aerospace Electronics*, June 18, 2010. http://www.militaryaerospace.com/articles/2010/06/ground-control-stations.html.
4. Y. Azoolai, Unmanned combat air vehicles, *GLOBES*, October 24, 2011, pp. 38–41.
5. S. Mraz, The military is beginning to rely on small unmanned drones or micro-air vehicles (MAVs) for scouting and reconnaissance missions. Dec. 9, 2009. MachineDesign.com.
6. N. Ungerleider, See what you can do with drone filmmaking. *Fast Company*, Jan. 31, 2013. http://www.fastcocreate.com/1682320/see-what-you-can-do-with-drone-filmmaking.
7. J. Hugen, Unmanned flights, *National Geography*, pp. 21–23, February 20, 2013.
8. BBC News Reporter, Drones to protect Nepal endangered species from poachers, June 20, 2012, pp. 13–15.
9. D.W. Casbeer et al., Cooperative forest fire surveillance using a team of small unmanned air vehicles, *International Journal of Systems and Science*, 37(6), 351–360, 2006.

2

UNMANNED AERIAL VEHICLES FOR MILITARY APPLICATIONS

Introduction

An unmanned aerial vehicle (UAV) is known by various names, such as a remotely piloted aircraft (RPA), an unattended air system (UAS), or simply a drone. Essentially, a UAV is considered an aircraft without a human pilot [1]. All aerodynamic functions can be controlled by onboard sensors, a human operator in the ground control location, or the deployment of autonomous electronic and electro-optical (EO) systems. The most basic functions of a military UAV include intelligence, reconnaissance, and surveillance (IRS). However, an unmanned combat air vehicle (UCAV) is supposed to meet combat-related functions in addition to IRS capabilities, such as target tracking and deployment of defensive and offensive weapon systems against targets. A civilian UAV can be equipped with simple electronic and physical sensors such a barometer, GPS receiver, and altimeter device. Sophisticated UAVs are equipped with photographic, television, infrared (IR), and acoustic equipment, compact synthetic aperture radar (SAR), LIDAR laser along with radiation, chemical and other special sensors to measure pertinent parameters to accomplish critical missions. Navigation and control sensors are of critical importance. Furthermore, the onboard sensors can be controlled by the ground-based operator, preprogrammed sensors, or automated remote operating mode. In the case of UCAV mode, mission requirements can be changed by the ground operator. UAV design scientists believe that when using the onboard and ground-based equipment, UAVs can perform a wide range of missions, such as intelligence gathering, surveillance, reconnaissance, aerial mapping, antiterrorist activities, and emergency operations with remarkable speed. Scientists further believe that the development of compact inertial navigation equipment, exotic software, and algorithmic maintenance for equipment calibration, filtering, and rapid and accurate processing of navigational information will enable UAV operators to perform important tasks with great accuracy and speed. Electrical design engineers are deeply involved in the specific development of onboard software and hardware of the next generation of computer vision and pattern recognition for navigation and UAV orientation. When the needed sensors and equipment are fully developed and available, UAVs can be equipped to create highly accurate images of the mouths of rivers, coastlines, ports, and settlements in critical regions. Typical physical parameters of a miniature UAV known as a nano- or micro-UAV, which can be launched by hand, are summarized as follows.

Typical physical parameters of mini-UAVs [1]

- *Takeoff weight*: 6–12 lb
- *Airframe weight*: 5–7 lb
- *Wing span*: 5–7 ft
- *Fuselage length*: 4–6 ft
- *UAV speed*: 20–30 mph
- *Payload*: 5–10 lb
- *Flight endurance*: 10–25 h
- *Rating of electrical motor*: 1 kW or 1.35 HP (some UAVs use gasoline engine and some use an electric motor)
- *Takeoff speed*: 15–20 mph
- *Landing speed*: 15–20 mph
- *Runway length*: 40–60 ft
- *Maximum climb speed*: 16 ft/s
- *Turn radius*: 35–50 ft
- *Flight altitude*: 50–5000 ft

Various Categories of Unmanned Vehicles for Combat Activities

There are various types of unmanned vehicles currently in operation. Unmanned vehicles can be characterized by the aerodynamic configuration of the vehicles and their operational functions. Some UAVs are designed to provide surveillance or reconnaissance functions and generally operate at low altitudes. Specially designed and developed unmanned vehicles are equipped with EO and electromagnetic sensors and weapon systems to provide combat capabilities. Note that combat UAVs generally operate at medium and high altitudes, while UAVs or drones generally fly at low altitudes ranging from 300 to 600 ft.

UAVs for Combat Operations

UAVs for military or combat operations are designed and configured to meet specific mechanical and structural requirements. Stringent structural, reliability, and stealthy features are given the highest considerations in the design of combat-based UAVs [2]. UAVs with conventional features are suitable to undertake IRS missions (Figure 2.1), UAVs with rotary-wing vehicle (URWV) configurations (Figure 2.2) are best suited for battlefield combat operations, and unmanned underwater vehicles (UUWVs), as shown in Figure 2.3, are ideal for undertaking underwater surveillance and reconnaissance of targets operating in coastal regions and tracking of hostile submarines operating under submerged conditions [3].

UAVs such as MQ-1 Predator (Figure 2.4) or MQ-9 Reaper (Figure 2.5) are recognized as hunter–killer UAVs. These UAVs are fully equipped with advanced navigation and communication systems, EO and electromagnetic sensors, and miniature offensive weapons to meet the stated combat mission requirements. The control of these combat vehicles can be accomplished either by a competent ground station operator or by an autonomous system deploying

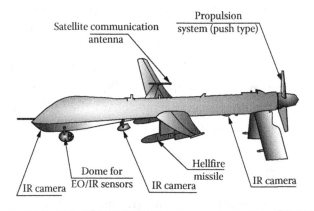

Figure 2.1   UAV platform for search, reconnaissance, and target tracking missions.

Figure 2.2   URWVs configurations are best suited for battlefield combat operations.

highly sophisticated speed computers, satellite communication system, and EO sensors.

Functional Capabilities of the GCS Operator

It is desirable to mention the capabilities and functions of the ground control station (GCS) operator. He is not just an electronic technician or a pilot. He is a combat-seasoned fighter–bomber who knows how to control the UCAV. The UAV pilot knows how and when to deploy the weapons aboard the UCAV using the sensors aboard the vehicle. The combat experience, EO sensor experience, and missile and weapon deployment capability of the ground station operator are the principal operational requirements. This operator must have full knowledge of the operating systems aboard the UAV and must be qualified to control the flight of the UCAV.

Description of GCS

The presence of the cockpit on the ground provides necessary help to the station operator to maneuver today's most sophisticated

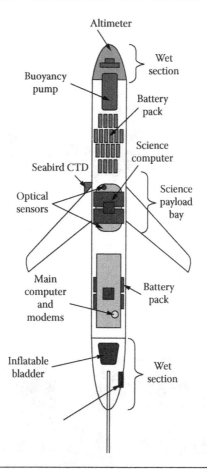

Figure 2.3   Unmanned underwater UAV for target tracking and tracking missions.

unmanned aircraft. Note that GCSs for UAVs or UCAVs are becoming a complex networking hub where the information from the UAVs is passed through two-way communications to the GCS and then to the whole battlefield network. Meanwhile, GCS designers and U.S. military officials are moving toward a common GCS to control all operating unmanned aircraft. Generally, a single operator is capable of handling a UAV or UCAV. However, in the case of a complex military vehicle, two control operators may be required depending on the complex equipment aboard the vehicle [4]. For example, the U.S. Navy MQ-88 Fire Scout unmanned helicopter is controlled by two operators in a room located aboard the mother ship [5].

Figure 2.4 Critical elements and EO/IR sensors associated with MQ-1 Predator UAV for missile-hunting missions.

Figure 2.5 Critical elements of the hunter–killer MQ-9 Predator equipped with multiple sensors and laser guided missiles.

The GCSs are generally located not only in or near the battlefield but also in controlled bases in the United States. A UAV controller or pilot could have breakfast with his kids and wife in the morning, head off to work, then fly combat missions over Afghanistan or Iraq, and ultimately head home for a family dinner at night. The following EO and electromagnetic sensors and displays or monitors are provided for the ground control operator and the UAV pilot on performance of their duties:

- Two consoles, one for the payload operator and one for the pilot or UAV operator
- Mouse and keyboard for the sophisticated onboard computers needed by the operators in the GCS
- Two moving map displays for each operator
- Two mission displays for the UAV pilot and payload operator
- Two-way secured phone lines for the pilot and payload operator
- Miscellaneous devices or sensors, which may help the GCS operators
- Two high-resolution displays
- Two monitors for the chat room

According to military experts, the GCS plays a central role in handling UAVs. As a matter of fact, the GCS acts like a hub for the IRS data collected by UAVs while operating over the regions of interest. For example, video and other data generated by various sensors such as SAR and EO sensors are downloaded via the secured data links to the GCS, and then that information is disseminated in real time for the troops in the battlefield and other agencies of interest. Note that the same groups can send this information to the GCS for upload to the UAC or UCAV for the aircraft or UAV to have the UAV fly to specific coordinates or make a strike on a new target obtained recently.

Based on the information available from the strike and surveillance program managers, a UAV GCS could have two consoles, one for the aircraft operator and the other for the payload operator [5]. Furthermore, the pilot or the operator sitting in the GCS is capable of controlling the UAV flight without a joystick. Instead, he commands changes in the vehicle flight by deploying a mouse and computer keyboard. Note the onboard computers are actually manipulating or controlling the control surface of the UAV to maintain vehicle speed and altitude at the desired values.

Note that the combat mission plans are preloaded in the vehicle computers, so it is conceivable that the UAV operator at the GCS can sit back and monitor his moving map display without interrupting the mission objectives. The level of autonomy or pilot interaction is dictated on a mission-to-mission basis. According to a combat UAV mission director, UAV operators are most often former fighter pilots or current pilots, and they are trained on air traffic control operations and flight dynamics under various climatic conditions.

A pilot generally uses two mission displays, a normal pilot cockpit display and a rotating or map display. Furthermore, the pilot has two other displays or monitors up and running displaying for multiple chat rooms. Note that the sensor operator also has two displays, one showing sensor status and the other displaying the route of the UAV flight. The sensor operator has two monitors for chat rooms just like the pilot. GCSs developed and operated by various Predator companies recommend some tasks by the aircraft operator or pilot and some for the payload operator. For example, the General Atomics GCS recommends that all the intelligence information coming from the IRS sensors, EO sensors, SAR, and systems should interface with others on the battlefield network. This scheme gives the pilot and payload operator complete situation awareness. The General Atomics Predator series GCS operators also use high-definition displays capable of providing additional information on the target of interest. In case of older UCAVs, the payload operator would be required to handle the basics of the computer system and to forward the incoming intelligence information from UCAV sensors to the intelligence analysts, command centers, troop commanders in the battlefield for their review and appropriate military action. The recent trend requires the payload operator to be trained as an intelligence analyst, so that he sees something important that needs to be investigated immediately and "alerts the troops in the battlefield and the analysis center for time-sensitive issues for immediate resolution. When something important occurs, the payload operator communicates immediately with the persons involved in the network via an online chat room, which is similar to the instant message services on the computers of the persons involved. In this case, the EO sensors aboard the UAV are tracking a terrorist or other critical target and this person may come in contact with a "red" vehicle or military police vehicle or other agencies may have been tracking. Under these conditions, the operator passes this information to others in the chat room for possible utilization in other military missions.

The payload operator is required to disseminate the high SAR images seen and the data collected by the data sensor to the wide area network (WAN) and intelligence analysts in other sectors or persons of interest who were looking for such information in a separate independent combat mission. Note that this is an example of distributed situational awareness, which significantly improves the battlefield picture for the UAV operators, troops on the ground or in battlefield, and intelligence analysts looking for information in real time. Some high-performance UAVs such as Predator B or Sky Warrior Alpha for deployment of Lynx advanced multichannel radar (AMR), designed and developed by General Atomics, are deployed to offer new information on dismounted targets.

Operating Requirements for UAV Operator or Pilot

To manage the AMR system, an operator is required to possess absolute working knowledge of the AMR, associated sensors and systems, GCS software, detection capability over its full field of regard (FOR), ground moving target (GMT) indicator formats, and high-resolution displays. It is important to mention that the GCS software is capable of supporting real-time cross-cueing to the EO/IR payload. The AMR system operator must be familiar with the very sophisticated performance requirements of the radar and its associated sensors. In summary, the Lynx AMR system provides the reconnaissance and surveillance capabilities over its full FOR and plug-and-play configuration for a high-performance UAV aircraft such as Predator B. It will be highly desirable to state that in case of UAV preplanned mission, the high-speed computer and software are capable of changing the aircraft location without interacting with the associated sensors. Reconnaissance and surveillance mission experts believe that both the

UAV aircraft and the mission operators are fully involved in the successful accomplishment of critical military missions.

Note that UAVs for law enforcement activities can be equipped with simple electronic and physical sensors such as a barometer, GPS receiver, altimeter sensor, and photographic, television, and acoustic sensors for monitoring the events on the ground. Radiation, chemical, and other special sensors are available for UAV applications, if required. As mentioned before, these sensors can be monitored or controlled by the ground operator in the GCS. Note that for commercial applications, generally few essential sensors or equipment such as communications and navigation are required for UAV, depending on the assignments involved. In case of geographical assignments, UAVs can be equipped with appropriate sensors capable of providing accurate images of river mouths, coast lines, sea ports, airports, and settlements in the critical regions of the country. In case of military UAVs, the constant flow of intelligence from long-endurance UAVs, such as the Global Hawk vehicle, enables the operators to feed the intelligence data to analysts and intelligence agencies around the world in near real time. This type of flow of intelligence data from one military agency to another is considered of significance combat importance.

Location of GCS

In combat operations, the location of the GCS is selected on the basis of military activities. However, the GCS location can be selected in a convenient location or in the vicinity of a military conflict area. In naval conflict situations, it is desirable to locate the GCS onboard the ship so that the MQ-8B Fire Scout unmanned helicopter has the GCS conveniently based aboard the ship rather than on the ground. The major difference between the GCS and that aboard the ship is that the shipboard communications equipment can be plugged in the ship's secured communication network, which can permit communications directly with the ship's air traffic control system via the internal communications systems. Furthermore, both the shipboard control station and the target illumination radar can be housed in the same room, thereby realizing substantial savings in cost and logistical operations.

Role of Portable UAV GCS

In the case of foreign military or civil conflict, a man-portable GCS can play a critical role. According to GCS designers, the controllers for small, man-portable UAVs are less complex, easy-to-use sensors or equipment that go from a system packed in a tight space on the war fighter's back to an operating system in less than 10 min. Under rapid change in political and conflict situations, a portable UAV GCS offers rapid removal and reestablishment from an old location to another new location, thereby saving significant cost and minimum interruption in UAV activities. Such was a case of removal of the GCS from a Pakistan territory to a border location of Afghanistan. One can see the benefits of this type of portable GCS when the bullets are flying around and the operators want to continue use of UAV operations for pressing military activities. Note that the operations of these portable GCS units do not need pilot training, because they are a part of small units of special forces operations, where everyone in the unit is trained for rapid responsibilities.

GCS designers believe that a family of small UAVs can have a common controller system comprising of various types of nano- and micro-UAVs, as shown in Figures 2.6 and 2.7. This controller system can control the operations of such small UAVs, which can significantly control the operating and personnel costs. A common

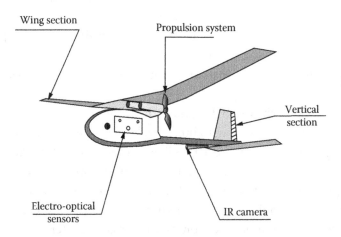

Figure 2.6 Micro-UAV RQ-11 B Raven vehicle provides a variety of military missions such as battlefield damage assessment, target detection, and conveyor security.

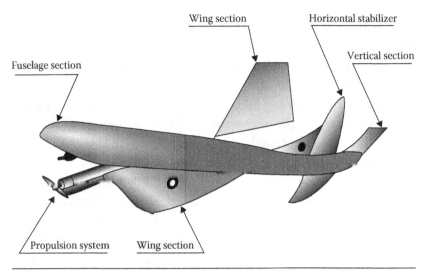

Figure 2.7 WASP III MINI-UAV best suited for reconnaissance and surveillance missions with minimum weight and size.

controller system is known as a portable modular system comprising a hand controller for launching the UAV, a compact and rugged laptop, and miniature transceiver unit, all plugged into a controller box, and the entire control is about the "size of two packs of cigarettes." This kind of box size is called a microcontroller. Note that this kind of laptop is available from Panasonic at moderate cost.

In such nano- and micro-UAV operations, one operator is needed with a controller module that consists of a display streaming video from the aircraft's cameras and miniature knobs and dials for controlling the UAVs. Another operator can be available if the laptop is required to collect the intelligence data downloaded from the aircraft for analysis and dissemination to the other nodes in the battlefield networks.

According to the GCS designers, the whole ground control system can be housed in the war fighter's back sack. The overall weight of such a portable GCS is close to 8 lb excluding the laptop. According to military packaging experts, one person can operate the system if the operation is not complex. However, a two-operator approach is essential for complex military missions. The miniature GCS can be embedded in a remote location as a command center that provides the same payload monitoring capabilities of operations in the field while performing some other functions.

Note that the duties and responsibilities of controllers for nano- and micro-UAVs are similar but do differ when it comes to payload control activities depending on the type of payload tool, which can vary from a mechanical gimbal sensor to an EO sensor while providing the same function. However, the GCS software recognizes which payload needs to be controlled and adapts to it. Combat UAV design engineers believe that the GCS could be operated by one well-experienced person, but the two-operator approach will be most reliable and practical. Because of complex and rapid thinking, ease of use of combat activities requires immediate attention in the design of a controller. It is further desirable to mention that constant attention paid by the operators on their displays could make them mentally tired and can create a cluttered view of the target. To provide ease of use to the operators, improvements should be made in the design of displays using decluttering technology. In addition, the display must be so designed to allow the individual operator to click on different windows as needed on the display screen.

Operator Responsibility for Payload Control

For old UAVs, the payload operator was trained to handle basic functions of the computer system and forward incoming intelligence information from the UAV vehicle to the intelligence analysts, command centers, and troops in the battlefield for detailed analysis. Now, there is a slight change introduced in the payload responsibility. The payload operator should be trained as an intelligence analyst, so if he sees something crucial or strange that needs to be exploited further as soon as possible and alerts the troops in the field as well as the analysis centers for time-sensitive targeting by saving time. This could be very critical. The payload operator communicates with other persons in the network through an online chat room, which is similar to common instant messaging services.

Role of Sensors aboard the UAV

EO sensors are best suited for tracking terrorists and other hostile targets in rugged hilly areas. These sensors also play critical roles in complex situations involving several targets. For example, the EO sensors of the UAV may be tracking the movements of a terrorist or other hostile targets through Kabul or Baghdad, while this particular terrorist may come in contact with a military police vehicle or other person that is being tracked by other agencies. Under these conditions, the operator passes this information to others in the same chat room such as the air operations center (AOC) for possible exploitations in other military search missions.

Images obtained and data collected by onboard EO sensors will be disseminated by the payload operator for possible transfer to WAN, and intelligence analysts in some agencies might recognize a person of interest or other target as data someone might have been looking for in an independent military mission. The images obtained from IR cameras and the data collected by various sensors are a form of distributed situational awareness that significantly improves the battlefield picture for UAV operators, troops on the ground, and intelligence analysts in near real time.

Role of Lynx Advanced Multichannel Radar

Most of the combat UAVs are not equipped with Lynx AMR systems because of high cost and complexity. Only some high-performance UCAVs such as Predator B and Sky Warrior Alpha aircraft are equipped with Lynx AMR because it is essential to get updated information on dismounted targets (which mean persons walking or running) for GCS operators and the battlefield network. Field tests performed on the AMR system detection performance over its full FOR using space time adaptive processing (STAP) indicated improved performance on discounted targets. Note that the GCS software supports real-time/cross-cue to the UAV aircraft's EO/IR payload. EO and IR sensors play critical roles in the detection of GMTs and GMT indicator formats, which are considered crucial in the detection of GMTs in the presence of ground clutter environments. The following EO/IR sensors and other elements are closely associated with the Lynx AMR system:

- Advanced multimode radar (AMR) with air-to-air and air-to-ground mode capabilities.
- Search, detection, and tracking of GMTs in clutter environments.

- Presentation of target information on high-resolution display for both the pilot and payload operator.
- Cross-cueing capability with the EO/IR elements.
- Tracking capability for dismounted targets on ground.
- Provision for tracking and identification of ground targets using high-resolution forward-looking infrared (FLIR) sensors.
- Radar also provides the emergency surveillance capability over its full FOR with minimum cost in high-performance UCAVs such as Predator B and Sky Warrior Alpha UAV.

The ability to detect and track dismount targets and slow-moving ground vehicles over large areas and to cross-cue the onboard video sensor to areas of significant interest is an emerging military and civilian surveillance requirement. The Lynx AMR provides this performance capability over its FOR with cost and plug-and-play configuration both for Predator and Sky Warrior Alpha UAV. As mentioned earlier, the GCS software supports the real-time cross-cue to the EO/IR payload, which provides improved detection of ground moving target (GMT) capability as well as GMT indicator formats. In brief, the pilot is able to see details of the GMT on his high-resolution display.

According to former U-2 pilots, both the UAV pilot and payload operators are much more involved in mission activities than ever. A complementary working relationship between them is extremely important if successful completion of the preplanned military mission is desired. This is because they both sit together in the same room of the GCS. The constant flow of intelligence information from long-endurance UAVs such as the Global Hawk permits the operators to feed it to analysts and intelligence agencies around the world in near real time. This information is of significant importance in military conflicts.

Locations of GCSs

Preliminary studies performed by the author on the subject concerned seem to indicate that for Navy UAVs and attack helicopters, the GCS should be ideally located on the navy ship or aircraft carrier deck, where needed logistic help and other technical support are readily available. For a complex and technically sophisticated

UCAV aircraft such as MQ-8B Fire Scout helicopter, a GCS can be installed on the ground as well as aboard the ship or aircraft carrier or on both. A major difference between the GCS and that aboard the ship is that the shipboard can be plugged into the communication networks, while the GCS has to deploy a separate communication system. The shipborne control system of the Scout helicopter is located in a room where a separate target illumination radar (STIR) is located. This illumination radar communicates with the air traffic control system of the ship via the internal communications system, while the GCS deploys a simple radio to communicate with the air traffic control system.

Critical elements and their functions of the MQ-88 B Fire Scout helicopter are briefly summarized as follows:

- *Performance capabilities*: Provided by the EO/IR system.
- *Sensor function*: Surveillance, reconnaissance, and target acquisition.
- *Function of laser designator payload provides*: Aforementioned three functions using illumination technology.
- *Target illumination laser radar*: Illuminates the target using laser energy.
- *Forward-looking infrared system*: This EO/IR laser designator payload provides surveillance, reconnaissance, and target acquisition functions at sea regardless of weather conditions. This particular sensor played a prominent role in the drug interdiction in the eastern Pacific region. This sensor has been successful in detecting and acquiring narcotics-operating boats in the same region, and the criminals were arrested by the law enforcement agents.

In summary, the sensors aboard the Fire Scout vehicle helped the agents to move in on the drug traffickers and seize the criminals and 60 kg of cocaine. Although it was not a military operation, the UAV technology has demonstrated the civilian uses in addition to military applications. The Fire Scout vehicle payload operator is capable of controlling different sensor functions such as laser range finder and communicates with the intelligence agents through the ship via secured radio communication and other communication systems carried by the ground forces.

Landing of Fire Scout Helicopter

Some UAV program managers believe that the most difficult task in flying an autonomous, unmanned helicopter at sea is landing the helicopter under windy conditions aboard a moving target or on a moving platform. Even after the end of its mission, the helicopter will hover behind the ship, wait for the signal from the ship to land on the ship deck, and deploy its sensors and instruments to determine the actual speed of the ship and its pitch, roll, and position to make a proper and safe landing on the ship's deck. One can now see that this particular helicopter, as shown in Figure 2.4, will face landing difficulty in unstable environments. To land the Fire Scout helicopter on the ship or aircraft deck, the helicopter pilot must be physically and mentally absolutely fit. Remember that the entire landing of the Fire Scout helicopter is monitored from the control station, which is located in the ship; in case any of the steps of landing malfunction, the landing can be aborted as soon as possible.

Deployment of Commercial-off-the-Shelf Components for the Control Station

Deployment of commercial-off-the-shield (COTS) components in noncritical subsystems can provide significant reduction in component as well as in system cost. For example, the Themis computer system (RES-32) in the Fire Scout control station uses the Sun Microsystems operating system, which uses COTS networking cards, input–output circuits, and peripherals. Using the aforementioned COTS components, one realizes a significant reduction in component cost as well as in system cost.

The racks used by the Fire Scout helicopter are qualified to meet military standard specifications. The software and protocols follow NATO standard specifications, which allow NATO members to participate in military operations with their own UAVs sharing UAV-generated [4] intelligence with each other, if required.

GCS for Each UAV Category

UAV program managers currently think that there are separate GCSs for each UAV category. It is a closed system. For example, a

Fire Scout helicopter–based GCS only works with a Fire Scout UAV and so on. This technique offers higher efficiency and confidence in the operation of the system. It is interesting to point out that sometimes there are differences within each platform depending on the end user or UAV program manager. For example, NASA's Global Hawk variant has different requirements compared to the Air Force UAV, and so these modifications in software programs can be made compatible with the performance requirements. Incidentally, the same is true for the European version of the Global Hawk known as Euro-Hawk.

Next Generation of GCS

In recent times, there has been an increased use of UAVs by various derivatives of GCSs. Calling out for commonality among the ground stations will significantly save money in training and development costs. The armed services are seriously considering the development requirements for GCSs, which will significantly cut the cost down. The Air Force and Navy call the GCSs as common ground segments. Essentially, it is a ground segment that is designed as an open-system architecture. These ground segments use the same latest high-speed computers, USB ports for peripherals, high-resolution displays, two-way secured communication equipment, and other sensors as deemed necessary. In brief, such stations will have a main operational system but different applications running to control different UAV platforms or EO and electromagnetic sensors. For example, a GCS is engaged in controlling the operation of a Global Hawk UAV and the associated team is about to complete the mission. Suddenly, a Fire Scout team comes into the same GCS and uses the station to fly the Fire Scout by switching to different software. The same station is now managing two different types of UAV platforms. This example illustrates the operational capability of the station operators using the sensors available in the GCS. Note that the system will be based on an open architecture but keeping the software independent from the hardware deployed. Based on the published technical article, Northrop Grumman Corporation has developed a GCS that works efficiently with all the company platforms. In the future, a particular military service wants a GCS that is strictly independent of both supplier and platform. Furthermore, an article published in *Military & Aerospace Electronics,* July 2010 reveals that General Atomics Company is actively engaged in the development of the next generation of GCS involving the advanced cockpit design concept that will significantly improve the situational awareness of the GCS pilot while reducing his workload. GCS designers believe that it definitely improves the situational awareness of the operator through the use of multiple high-definition displays that deploy digital elevation and train data concepts to permit a 120° view across the large high-definition screens in the cockpit. Using aforementioned concepts, computer symbols can be electronically presented on the high-definition screens, which will provide the pilot a comprehensive common operating picture of the aerospace in which the UAV pilot is operating. It is further important to mention that the 120° view across the high-definition display screens augments the narrow field of view (FOV) of the camera system. Essentially, the advanced cockpit design optimizes the flight crew performance in crowded space environments. Note that the advanced cockpit design uses intuitive touch screen displays to reduce significantly the training time and cost. The improved cockpit performance is strictly due to sophisticated software because of the use of three-dimensional moving maps capable of providing the pilot with a unique ability to track accurately the Predator series UAV's movement in a multidimensional display. Furthermore, the software also employs a "modular open architecture that permits rapid incorporation of local-made components, which not only reduce the costs but also significantly improve the capability and interoperability."

The advanced cockpit is complete and the flight tests are about to be concluded. The advanced cockpit will be available for deployment by the UAVs. The commonality problem will be solved through the sophisticated software available. Attempts will be made to use COTS components as much as possible depending on the mission criticality. COTS program managers believe that high-density COTS I/O solutions reduce board count, package volume, weight, and power consumption, which will significantly enhance overall system reliability and affordability. Common standards, COTS components, open architectures, and calibrated latest equipment will accomplish the mission objectives.

As the military services move to the next-generation common ground station, they will be testing the performance capabilities of the EO and electromagnetic sensors associated with the UAVs. Note that determination of the type of system to use is strictly dependent how one defines the real time. For example, the time taken for the Fire Scout UAV operators to initiate a command and see on the wide screen if that is carried out by the aircraft needs to be short enough to be instantaneous to them. Furthermore, it needs to be deterministic but not necessarily strictly real-time critical in some radar or missile system.

Note that the technology to accomplish the issue is here, but it takes time to settle the requirements between various military branches. Interesting enough is to mention that both the Air Force and Navy are working closely to make sure that the Global Hawk and the Bay Area Maritime Surveillance (BAMS) systems are closely aligned and are compatible in the open architecture for ground control system.

Impact of Human Factors on Control Station

GCS operators provide a lot of operational requirements to the control station contractors in designing the next-generation control stations. The operator recommends a larger display screen that can replace two screens currently: one for the aircraft operator or pilot and the other for the payload operator in the Fire Scout aircraft. Note that the larger screen dimensions are 16 × 9 in, and it is best suited for a high-resolution display. Seasoned UAV operators prefer the large screens and window configurations that accommodate the individual tastes of the aircraft operator and mission payload operator. In the past, military services would provide one fixed display configuration for the pilot and the payload operator, which really reduces the efficiency of the operators as well as their comfort in the narrow space available in the control station. Many young engineers coming out of colleges and boot camps are used to the flexibility of modern personal computers and large video game screens, and they work efficiently even over long hours.

Weapons Best Suited for High-Value Targets

It is highly desirable to provide light deadly weapons such as the Stinger missile or Hellfire missile or any other light deadly weapon to the UCAV (UAB) operators. These UAV operators are very sharp and quick and are capable of deploying an appropriate weapon against the enemy radar or fighter aircraft. These offensive weapon systems must be light, deadly, and offer minimum undesirable aerodynamic forces to the UCAV flight.

Combat UAVs Operated by Various Countries [1]

The author wishes to summarize the performance capabilities of UCAVs operational around the world and owned by various countries. The ADCOM (Arab Defense command) system comprising 40 UAVs has been developed for the United Arab Emirates to support various operations in the Arab regions. Its basic missions include intelligence, reconnaissance, and communication (IRC) relay platform capable of supporting specific support operations such as special forces, regional army divisions, humanitarian aid responses, and damage assessment. According to military experts, this system can be easily modified for combat roles such as carrying 100 kg missile pods leading to a total payload close to 220 kg. The UAV has a serpentine like fuselage carrying a single vertical tail with no planes. A three-blade propeller located at the extreme rear end and arranged in "push-configuration" intakes can be seen at either side of the fuselage aft. This UCAV is powered by twin-engine arrangement with its primary engine installation being a conventional unit rated at 115 HP. This is coupled with an electric system with a rating of 80 HP. Critical design parameters of various UCAVs can be described as follows [1]:

United 40: Emirate

- *Speed limits*: 75–200 km/h
- *UCAV length*: 11 m
- *Wing span*: 4.4 m
- *Vehicle empty weight*: 1150 lb
- *Maximum takeoff weight*: 3300 lb
- *Combat ceiling*: 7000 m
- *Mission endurance time*: 120 h
- *Year of service*: 2016
- *Crew*: 0

BAe UCAV: European UAV

BAe Systems has developed two unique combat vehicles, namely, Mantis and Taranis. They hailed the Mantis strictly as a UAV technology demonstrator to evaluate a full-size UAV. This UCAV supports the delivery of precision-guided munitions as demonstrated by MQ-9 Reaper family. Note that these UCAVs have large dimensions to accommodate more advanced in-flight weapon systems and fuel capacity to meet longer operating range and mission payloads.

The Mantis program intends to produce a reusable, long-range, deep-penetrating UCAV. This particular UCAV is designed for high endurance, long-range performance capability, and modeling payload capacity including weapon delivery. This vehicle is designed to operate on largely autonomous systems by relying on satellite communication for its self-positioning capability. Note that Mantis is independent in its navigation, landing, and takeoff procedures. The Mantis development program was headed by BAe Systems with critical support from GE Aviation, L³ Wescom, and Rolls-Royce. The electrical system includes a high-resolution imaging sensor and imagery-collection and exploitation (ICE) system.

The profile is largely conventional and similar to a manned aircraft. This domed front section houses all EO and mechanical sensors. The fuselage is well contoured for high aerodynamic efficiency. The rear section is tapered and mounted with a "T"-shaped tail unit. The undercarriage is wheeled in the traditional wall and is fully retractable. The wall is designed to support external munitions across six hard points. Estimated physical parameters can be summarized as follows:

- *Overall vehicle length*: 65 ft
- *Aircraft empty weight*: 2200 lb
- *Takeoff weight*: 19,800 lb
- *Power plant*: Two Rolls-Royce series turboshaft engines, each capable of developing 380 HP
- *Top speed*: 345 mph
- *Cruise speed*: 230 mph
- *Mission endurance time*: 30 h
- *Source*: www.militaryfactory.com
- *Maximum range*: 9942 mi

- *Service ceiling*: Not available
- *Rate of climb*: Not available
- *Armament suite*: Variable precision-guided munitions

BAe System Taranis: British UAV [1]

The BAe Taranis offers effectively bridges the gap between the original UCAV of yesterday and the full-size unmanned jet-powered multirole aircraft of tomorrow. The Taranis is a UK developmental unmanned system. This is a completely autonomous vehicle and is capable of precision guidance munitions delivery at long operating range. This is not a military end product at present. It is important to mention that it is capable of preprogrammed waypoint following, takeoff, and landing procedures by making in-flight decisions with ground control operators override.

The vehicle size is close to that of BAe Hawk jet strike aircraft with low-observable stealth capability. The aircraft wing design is aerodynamically efficient and the structure is well contoured. The physical parameters of the Taranis are summarized as follows:

- *Wing span*: 30 ft
- *Ground height*: 30 ft
- *Aircraft weight*: 18,000 lb
- *Power plant*: Turbofan engine with output thrust around 6500 lb
- *Flight capability*: Supersonic
- *Flight test status*: Tests completed in August 2013 in southern Australia
- *Source of information*: 2003–2014. www.militaryfactory.com
- *Maximum operational range*: Not available
- *Service ceiling*: Not available
- *Rate of climb*: Not available

Dassault nEUROn (European UCAV)

This UCAV is developed by Dassault, France. It is a joint European venture that includes various countries, including Sweden, Italy, Spain, France, Greece, and Switzerland. Note that Dassault is responsible for the final assembly and in-flight testing of the aircraft. Furthermore, the weapon suite will be handled by an Italian corporation while the communication system will be provided by Spain and France jointly. The aircraft has been given a conventional delta-wing shape with all protrusion managed along the upper wing surfaces. It has a single engine installation at the center-top of the design with a trapezoidal air intake to reduce IR signature, and the exhaust structure is designed with a low radar cross section (RCS). The aircraft is designed with stealth features and there are no vertical tail surfaces, thereby making the nEUROn a "flying wing" design. The undercarriage will be fully retractable into the fuselage underside while the intended weapons bay will be internally held to provide additional stealth capability. Physical parameters of the vehicle are as follows:

- *Aircraft length*: 31 ft
- *Wing span*: 41 ft thus putting it on par with a full-size main aircraft design or UCAV system similar to Lockheed RQ-170 Sentinel aircraft
- *Empty aircraft weight*: 10,800 lb
- *Fully loaded UCAV aircraft weight or full-mission load*: 15,400 lb
- *Maximum speed*: 600 mph (supersonic speed)
- *Power plant*: Single Rolls-Royce turbofan engine centrally located in the fuselage
- *Service ceiling*: 46,000 ft
- *Flight control and weapons delivery design*: With full autonomy in mind and human interaction
- *Source*: 2003–2014. www.militaryfactory.com

Rustom (Warrior): Indian UAV [1]

This particular vehicle is developed and managed by the Defense Research and Development Organization (DRDO) of India. The DRDO is working on the Rustom UAV program with medium-altitude long-endurance (MALE) capability with prospects for an armed version ultimately. The program initiatives were taken in the 1980s and 1990s, which resulted in a flyable prototype in November 2009. The Rustom will have three distinct variants, namely, Rustom-1, the high-altitude long-endurance (HALE) vehicle of 12 hours, Rustom-2 with flying altitude close to twice the altitude of Rustom-1 and Rustom-H. According to DRDO scientists, Rustom-2 will be a fully featured combat vehicle with functional capability similar to that of the American Predator UAV line. The system sports a three-legged, fixed undercarriage, rear-mounted main wing appendages, and a canard wing assembly at front. Vertical fins are set at the main wing tips. The power plant includes a two-blade propeller in a "push-up" configuration and is located in the aft section of the fuselage. Avionics and mission electronics are housed within the forward section of the fuselage. Physical parameters of the Indian UCAV (Rustom-1) can be summarized as follows:

- *Maximum payload*: 165 lb
- *Empty vehicle weight*: 1560 lb
- *Power plant*: American Lycoming 0–320 four-cylinder engine with 150 HP rating

Parameters for Rustom-2 (Indian UAV)

- *Payload limit*: 770 lb.
- *Empty aircraft weight*: 4000 lb.
- *Power plant consists of* two Russian NPO Satwas 36 MT series turboprop engines.
- *Engine location*: Under the wings in nacelles.
- *Each engine power output*: 100 HP while driving a three-blade propeller.

Parameters for Rustom-H (Indian UAV)

- *Flight scheduled for Rustom*: 2014
- *Maximum speed*: 140 mph
- *Operating range*
  - 140 mi (Rustom-1)
  - 220 mi (Rustom-2)
  - 625 mi (Rustom-H)
- *Maximum operating altitude*
  - 15,000 ft (Rustom-1)
  - 26,000 ft (Rustom-2)
  - 35,000 ft (Rustom-H)

Remarks: Rustom-2 is intended to replace the Israeli IAI Heron line of UAV, which is currently deployed by Indian Air Force and Indian Navy.

- *Endurance limit*
  - 12 h (Rustom-1)
  - 20 h (Rustom-2
  - 24 h (Rustom-H)

Note: Rustom-1 and Rustom-H are not designed as *armed* UCAVs and hence, are not suitable for combat roles.

Israeli UAVs

Elbit Hermes 450 vehicles are being deployed for communication, reconnaissance, and surveillance missions. Hermes 450 systems are extremely popular, and these UAVs have been purchased or leased by several countries, including the United States, the United Kingdom, India, Singapore, Brazil, Croatia, Mexico, and others. The Hermes 450 vehicle is a UAV designed, developed, and evaluated by an Israeli company. Several versions or variants of this UAV have been manufactured. Note that Hermes 450 falls under the Elbit Systems Corporation and is specifically designed for tactical long-endurance sorties. This vehicle is primarily utilized as a communication, reconnaissance, and surveillance platform by the Israeli Air Force (IAF). An armed version or variant of Hermes 450 is thought to exist and is operational in India and other foreign countries [1].

It is interesting to mention that Elbit has designed and developed the Hermes GCS, which allows the flight crew located in a GCS to have full control over all integrated Hermes 450 systems. This arrangement allows to interpret all incoming imagery and data in real time and to take appropriate action and more often in conjunction with other ground control elements. The Hermes has demonstrated a proven and highly versatile defense system that can be modified to suit a specific tactical mission. This system deploys an EO payload that can accept the reconnaissance and surveillance sensors as needed to meet the mission requirements. The Hermes manufacturing company Elbit states that their Hermes 450 system has accumulated more than 65,000 hours of flight in its short history and has been in active service in several combat environments.

As far as the structural integrity is concerned, the Hermes 450 strictly follows in line with other contemporary UAV designs and shares an external shape similar to the American Predator series. The fuselage is long and has a tubular shape, which contains the required guidance system, payload suites, and internal guidance controls. The undercarriage consists of a bow-legged structure mounted amidships holding single-wheeled main landing gear legs, while the steerable nose-landing gear involving a single wheel is held well ahead in the vehicle design, just aft of the nose cone. Note that the undercarriage is completely static and does not retract during the aircraft flight. The EO suite can be seen protruding from its blister housing along the fuselage centerline ahead of the vehicle wings. The wing assembly sits atop a short-support, rigid structure and is straight in design with cranked-wing tips. The aircraft wings sit just aft of the amidships. The empennage is made of a "V-shaped" slanted wing structure and its wings doubling as both the rudder and stabilizer. The power plant consists of a compact, single, Wankel four-stroke rotary engine. The power plant meets the necessary operating range and endurance requirements to the Hermes 450.

Estimated parameter values for Hermes 450

- *Fuselage length*: 20 ft
- *Wing span*: 34.5 ft
- *Operating range*: 125 mi
- *Gross weight*: 1000 lb including empty aircraft weight
- *Endurance capability*: 20 h
- *Rate of climb*: 900 ft/min
- *Maximum speed*: 110 mph
- *Cruise speed*: 81 mph
- *Weapon suite*: Two Hellfire antitank missiles

UAVs Operational in the United States

A comprehensive review of published technical articles [1] and reports seems to indicate that the United States has designed and developed state-of-the-art UAVs using advanced technologies deploying the autonomous concept and state-of-the art sensors. Note that the Predator and Reaper series represent the most advanced UAVs and UCAVs. Furthermore, the UCAVs or UAVs operated by the U.S. Air Force, Navy, or Army might have different operational, endurance, speed, ammunitions, and sensor requirements depending on the mission objectives. Even though the UAVs are mostly deployed for military applications, their usefulness is reaching beyond the battlefield such as disaster relief, emergency medical assistance, and homeland security operations.

MQ-1 Predator Series This particular UAV is used as an unmanned reconnaissance platform and as a missile hunter [3]. The MQ-1 is designed by remote control technology and covers a territory with excellent loitering endurance without exposing to enemy fire or capture. The Predator is operated remotely by three experienced operators, including one pilot and two EO sensor operators. The vehicle is controlled and operated by the operators sitting in the GCS, which is fed by ground equipment and a satellite communication link. One full Predator group consists of four Predators. The Predator design requires very little surface area for landing and takeoff. The landing is accomplished with a retractable tricycle landing gear system. The flying function is accomplished by the pilot sitting in the GCS and using a joystick and forward-mounted color camera. Note that the IR and TV cameras are located in the fuselage to provide real-time and image recombination capability. In case of armed reconnaissance mission, Hellfire antitank missiles are available for combat operation and vehicle safety.

General Atomics MQ-9 Reaper This MQ-9 system is also known as Predator B UAV. The MQ-9 Reaper or Predator B serves as the hunter–killer for the USAF. The symbol "M" stands for multimode, "Q" for UAV, and "9" for series designation. Note that the MQ-9 Reaper or Predator B offers the USAF a high-level, remotely piloted weapon delivery platform capable of providing instant offensive response and precise engagement. Essentially, the Reaper is a large derivation of the Predator series of UAV and features more power in terms of delivering capabilities of both power plot munitions. Furthermore, this platform offers high-quality images of the targets due to use of intensified TV, daylight TV, and IR camera along with an integrated laser range finder. The MQ-9 will be superseded by a larger, jet-powered, stealthy Avenger series (Predator C) in the near future.

Physical parameters and their estimated values for MQ-9 Reaper:

- *Wing length*: 36 ft
- *Wing width*: 66 ft
- *Vehicle height*: 36 ft
- *Empty vehicle weight*: 3695 lb
- *Fully loaded vehicle weight including fuel*: 10,500 lb
- *Power plant*: One Honeywell turboprop engine with 900 HP rating
- *Maximum speed*: 230 mph
- *Maximum operating range*: 1878 mi
- *Maximum service ceiling*: 50,000 ft
- *Number of hard points*: Four
- *Armament suite consists of the following*:
  - Two Hellfire antitank missiles (AGM-114)
  - 2 GBU-12 or 2 GBU-8 joint direct attack munitions
- *Current operating countries*
  - United States
  - United Kingdom
  - France
  - Italy
  - The Netherlands

Guizhou Sparrow Hawk II (Chinese UAV)

The author suspects that the design shape of the Chinese Sparrow Hawk II is close to the American General Atomics Predator UAV. This Chinese UAV is intended for IRS missions. Its payload includes advanced EO systems and IR sensors. This aircraft gets its power from a single engine driving a three-blade propeller in a "pushing" configuration. Its engine is as in the Predator UAV and located at the rear of the fuselage. The fuselage is slender and is well concaved for a flight with a blister pack mounted under the fuselage for its payload-carrying equipment. The undercarriage has three wheels in triangle configuration to allow the UAV to land and take off from the prepared runway.

The Sparrow Hawk II UAV completed its flight tests in August 2011 and the flight lasted for 4.5 hours. This UAV has been photographed with under-wing hard points, and dummy missiles demonstrated the Hawk II as a UCAV in the near future Parameters and their values for Sparrow Hawk II can be summarized as follows:

- *Dimension of wing*: Not available
- *Power plant*: One rear-mounted engine driving a three-blade propeller in "push" configuration
- *Maximum speed*: Not available
- *Maximum range*: 1800 mi
- *Maximum service ceiling*: 28,000 ft
- *Armament suite*: Air-to-surface guided missiles: Optical and IR sensors payload in drop/launch ordinance
- *Source*: MF network: GlobalFirePower.com

Guizhou Soar Eagle Chinese UAV

Based on the aircraft imagery, the Soar Eagle UAV bears a distinct resemblance to the Northrop Grumman Global Hawk long-range, high-altitude UAV system of the USAF. Essentially, the Soar Eagle is a HALE UAV. But the Chinese Soar Eagle design includes a single vertical tail fin and a joint tendon wing configuration. The Soar Eagle, like the Global Hawk, has a large fuselage close to the size of a small aircraft. This aircraft is specially designed for long-range, high-altitude mission, which is stated as HALE UAV. The Soviet-developed engine with a thrust capacity of 9800 lb is widely used in the MIG-21 fighter and SU-15 interceptor. This engine was designed and developed at the end of the 1950s and suffers from high failure rates. Published reports reveal that the Indian Air Force lost over 100 MIG-21 fighters due to a combination of engine failures and pilot errors.

Physical parameters and their estimated values for Chinese Soar Eagle

- *Wing span*: 81 ft
- *Fuselage length*: 49 ft
- *Aircraft height*: 18 ft
- *Maximum operating range*: 3495 mi
- *Power plant*: WP-13 turbojet engine (which is a true copy of the Soviet-era Tumansky R-13 series jet)
- *Output thrust*: 9700 lb
- *Endurance window*: 10 h
- *Cruise speed*: 460 mph
- *Armaments*: Specifics not available
- *Basic capability*: Limited to reconnaissance sorties only
- *Source*: MF network: GlobalFirePower.com

Miscellaneous UAVs Designed and Developed by U.S. Companies

So far, the author has described the most significant and high-performance UAVs operated by the United States, French, Israel, China, and the United Kingdom. Now, the author wants to focus on the low-capability UAVs, micro-UAVs, nano-UAVs, and flapping wing UAVs. UAVs designed and developed by the Israeli company Elisra, U.S. Naval Research Lab (NRL), AAI Corporation (USA), AeroVironment of Simi Valley (USA), and other research organizations will be briefly described with emphasis on performance capabilities, cost, and critical application.

Smallest UAV Developed by NRL (USA)

Among the smallest UAVs, the Dragon Eye UAV designed and developed by the NRL has been used by U.S. Marines in Iraq by force-protection troops to search the waters ahead of a submarine for threats. The compact Dragon Eye UAV weighs only 5 lb and can be disassembled into *five* pieces within less than 5 min and can be stored in a compact suitcase, which can be carried by hand. This UAV has been developed by the NRL as a stealth information-gathering tool. The Dragon deploys a quiet electric motor that runs about 50 min flying time on a single lithium-ion battery charge and can achieve a range of 40 km or 128,000 ft and an altitude close to 10,000 ft. Smaller UAVs are essentially gathering information and sending data back by means of a short-range line-of-sight (LOS) microwave link. Note that many microwave links are based on spectrally inefficient analog modulation techniques, such as frequency modulation (FM). If too many such UAV systems are operating at the same time, both the forward signals and the return data are lost. Both of these problems can be eliminated by using a time-division-duplex (TDD) technique to achieve a spectral efficiency of 4 MHz per channel. Furthermore, this technique can work on single-channel or frequency-hopping modes. Note that the data links connect UAVs to operators and control stations at distances to 60 mi.

U.S. UAVs for Space Applications

Currently, the main focus on UAVs is on tactical applications. Some companies as well as government agencies such as the NASA are exploring other uses of the unmanned remotely controlled vehicles. NASA has been working with General Atomics Aeronautical Systems in San Diego as a part of the Environmental Research Aircraft and Sensor Technology (ERAST) program at Edwards Air Force Base, CA, to design and develop a modified version of Predator B UAV for high-altitude missions. The extended wing span design is best suited for high-altitude earth missions at altitudes greater than 65,000 ft. Defense officials believe that such UAVs can be deployed for military applications such as emergency food and medical supplies when needed and in restricted areas where regular aircraft landing is not possible. These high-altitude UAVs are most suitable for applications such as disaster relief and for border patrol as a part of the Department of Homeland Security (DDS) operations.

Classification of Small UAVs

The number and types of unattended aerial vehicles have grown considerably since their humble beginning. These UAVs can be categorized by size or tier. Note that Tier-1 represents low-altitude UAVs, Tier-2 indicates MALE vehicles, and Tier-3 represents HALE vehicles. Furthermore, different branches of the armed forces deploy different aircraft, with different models falling under different tier categories [6]. Now some specific UAVs will be described briefly with emphasis on performance capabilities and limitations.

RQ-7 Shadow UAV Developed by AAI Corporation (USA)

This particular UAV was designed and developed by the AAI Corporation of the United States. The Shadow UAVs (RQ-7) is used by both the U.S. Army and U.S. Marine Corps. This vehicle can be launched as trailer-mounted pneumatic catapult. The UVA is equipped with a liquid-nitrogen-cooled EO IR camera and C-band (5250 MHz) LOS data link to carry high-resolution images and real-time video to a UAV control station.

It is interesting to point out that the smaller Tier-1 UAVs include the Desert Hawk, designed and developed by Lockheed Martin Corp and Raven UAV aircraft, designed and developed by AeroVironment of Simi Valley (CA). These particular small UAVs can be launched by hand if the weight permits. The Tier-2 UAVs are relatively larger vehicles and include the Scan Angle from Insitu (www.insitu.com), the Shadow UAV from AAI Corp, and the Sentry UAV from DRS Technologies [6]. The largest UAVs fall under the Tier-3 category, and they include Global Hawk UAV developed by Northrop Grumman Corporation, the rotary-wing Fire Scout from Boeing Company, and the Predator from General Atomic Company in San Diego.

Electronic circuit technology and sensor technology must be given serious consideration. Standard electronic devices and sensors may not be suitable for the deployment particularly in Tire-1 and Tire-2 category UAVs. Improvements in weight, size, and power consumption may be necessary for their use in Tier-1 and Tier-2 categories of UAVs [6].

UAV for Maritime Surveillance [7]

The Northrop Grumman Corporation–built UAV provides the most compact platform best suited for maritime surveillance missions by the U.S. Navy. This Broad Area Maritime Surveillance Demonstration

(BAMS-D) UAV is flying currently 15 missions a month and allows fleet commanders to identify and track potential targets. This particular platform provides a strategic picture to the carriers and amphibious battle groups as they move from area to area. Important performance parameters can be summarized as follows:

- *Endurance capability*: 24 h
- *Operating altitude*: 25,000 ft (minimum)
- *Sensors capability to monitor large ocean areas*: Over 360° FOV
- *Maintenance capability*: Anti-ice and de-ice function
- *Structural strength improvements*: Provided to operate under rough weather conditions

Miniaturized Components for Synthetic Aperture Radars

Studies on component requirements performed by the author indicate that conventional EO, electromagnetic, and electronic components may not able to fit into the space available in the Tier-1 UAVs. Furthermore, Tier-1 UAVs would prefer minicomponents with minimum weight, size, and power consumption. The author has made a component survey for possible applications in Tier-1 and Tier-2 UAVs. The following critical components have been summarized for Tier-1 and Tier-2 UAV applications including shock-proof and vibration-proof microconnectors for optimum reliability in harsh aerodynamic environments.

Miniature Sensors for Reconnaissance Missions by Small UAVs

Uncooled FLIR sensors are best suited for reconnaissance missions undertaken by Tier-1 and Tier-2 UAVs. EO sensor experts believe that the uncooled FLIR sensors will give soldiers, sailors, airmen, marines, and law enforcement agents a powerful tool in conducting reconnaissance activities with minimum cost, size, and weight. This low-cost, lightweight, high-performance sensor can be either handheld or mounted under the UAV nose or belly. This FLIR is so small that it can be used as a pocket scope. Two expensive components of uncooled FLIR sensors are the detector and the optics. Reconnaissance managers claim that for a UAV sensor that needs a fairly wide FOV and a short-range performance, uncooled FLIR is found the most cost-effective EO sensor for reconnaissance missions. In brief, the uncooled FLIR is best suited for short-range applications. Highlights of this FLIR sensor can be summarized as follows:

- *Maximum weight*: 14 oz
- *Run time*: 3 hours (maximum)
- *Human detection range*: 500 m or 1600 ft, approximately
- *Zoom capability*: 4× while providing the human detection range

Uncooled Thermal Imaging Camera for Small UAVs

DRS Sensors and Targeting Systems, Inc, Dallas (Texas), offers uncooled-microbolometer-based thermal imaging camera engines for electro-optic-mechanical (OEM) integration on small UAVs. These engines include an uncooled detector, system electronics, housing, and lens. Close examination of this imaging camera indicates that the DRS E3500S model split electronics configuration camera engine packs plenty of performance in a very compact and flexible interface design. It has low input power requirement and its performance includes the 320 by 240 pixel, 25 μm pixel uncooled microbolometer detector. Note that the DRS E6000 offers user-programmable digital signal processing (DSP) technology and rugged housing with high mechanical integrity, improved sensitivity, and low-power requirement. Outstanding features of this imaging camera can be summarized as follows:

- *Detector resolution*: 640-by-480 μm
- *Pixel pitch*: 25 μm
- *Weight of thermal imaging camera*: 5.8 oz (maximum)
- *Nominal power consumption*: 3 W

Miniature Synthetic Aperture Radar Surveillance

Compact side-looking radars (SARs) or miniature SARs have been designed and developed for UAVs to undertake surveillance and reconnaissance missions. These SARs are available for deployment in very small or micro-UAVs. To compensate for larger SAR dimensions, the most efficient signal processing technology involving the latest signal processing algorithms is used. Tomography formulation of the spotlight mode of SAR is given serious consideration, which provides improved angular resolution. The pulse compression technique involving efficient linear FM-matched filter algorithms is deployed for SAR signal processing to achieve enhanced range resolution. Published articles on micro-SARs reveal that the weight of SARs is close to 5 lb with minimum size.

Miscellaneous Compact Sensors for Tier-1 and Tier-2 UAVs

1. Stealth computer
   a. *Stealth computer*: Offers stealth features
   b. *Size*: Small enough to fit in the palm of the hand
   c. *Core size*: 2-dual processors (close to the size of a hardcover novel)
2. Forward- and side-looking IR cameras
   a. *Maximum weight*: 6.5 oz
   b. Detection range for a man-size target, 820 ft during day and 320 ft at night
   c. Switching time between an EO sensor and IR camera: 15 s
3. Miniaturized batteries for UAVs
   a. *Battery chemistry*: Relies on electrochemical reaction to provide electrical energy
   b. *Service life*
      i. 7–10 years (alkaline battery)
      ii. 10–15 years (lithium battery)
      iii. 20 years or longer (lithium-thionyl-chloride battery)

Data Link Types

- *Type*: analog; digital
- *Operating RF band*: C-band LOS link for minimum payload
- *Modulation techniques available*: FM or TDD

Note that the TDD modulation technique offers a spectral efficiency of 4 MHz per channel and can work in single-channel or frequency-hopping modes. The data link connects the UAVs to operators and control stations at distances not exceeding 60 mi. It is interesting to point out that EnerLinks II full-duplex data links from EnerDyne Technologies provide digital over analog FM links for a number of UAV platforms. The performance of a data link is strictly a function of transmitter power, power amplifier AM/PM characteristics, and RF antenna band. Optimization of the aforementioned parameters is essential to achieve high-quality compressed video and network data traffic with as much as 12 dB improvement in the link margin over the existing analog link system.

NANOSAR-C

The author has reviewed the latest technical reports and articles on NANO-SAR sensors for possible applications in Tier-1 and Tier-2 unattended aerial vehicles. Comprehensive review of NANO-SARs and MICRO-SARs using NANO and MICRO technologies is most ideal for small UAVs. Particularly, a NANO-SAR operating in the C-band provides cost-effective detection, location, and classification of GMTs in rain, snow, fog, dust, smoke, and day/night environments where other sensors fail. The NANO-SAR-C sensor is the world's smallest SAR and light enough to be mounted in Tier-1 and Tier-2 UAV aircraft. NANO-SAR-C is integrated with Lisa ground station and Viper™ communication link to provide a cost-effective plug-and-play radar imaging solution as well as real-time monitoring and control of NANOSAR-C radar and real-time SAR data analysis. The payload includes the SAR radar, turret, RF antenna, and cabling with microconnectors mounted either in a 7-in. diameter wing-mounted pod or inside the aircraft fuselage. Even some space is available to accommodate other sensor payloads needed for extended system endurance capability.

Operating Modes

- Strip map
- Spotlight
- Circular
- Moving target indicator (MTI)
- Command and control (provided by Lisa 3D and Lisa Dashboard)

- Communication provided by TTL, RS-232, Ethernet
- Sensor cueing by using a cursor on target
- Image products available from Google Earth, Complex NITF, JPEG, PNG, and BMP

Image Processing and Exploitation

- Lisa image offers real-time image processing.
- Lisa change provides change detection.
- Lisa three-dimensional provides image exploitation, control, and flight planning.

System Performance Parameters

- *Transmitter power*: 1 W
- *Frequency band*: K_u, X, UWB, UHF (subject to approval by the customer and supplier)
- *Range resolution meter (ft)*: 0.3 (1), 0.5 (1.565), 1 (3.28), 2 (6.56), 5 (16.5), and 10 (32.8)
- *Standoff range*: 1–3 km
- *Slant swath*: 2000–4000 resolution cells, which means 2000–4000 ft at 1 ft resolution
- *Power consumption*: 25 W (maximum)
- *Sensor weight*: 2.6 lb
- *DC Supply voltage*: 12–28 V
- *Radar system size*: 5.5 × 3.5 × 2 in including RF antenna and IMU

Options Available

- Extended range possible with detection technology change.
- Dual polarization/interferometry offers MTI capability.

Hunter–Killer UAVs for Battlefield Applications

The crowning success of hunter–killer UAVs in recent battlefields has convinced the defense officials and military planners that future military conflicts are strictly dependent on the next generation of such UAVs. Tactical experts and UAV officials firmly believe that increased weapon capability and a weaponized intelligence, surveillance, and reconnaissance (ISR) platform are vital in the battlefield. The first operational Reaper MQ-9 arrived at the Creech Air Force Base in Nevada and started laying the groundwork for hunter–killer squadron operation for the USAF [3]. It should be remembered that MQ-9 is a multimode vehicle and is fully equipped with advanced weapons and sensors. Operations of these sophisticated UAV are handled by well-experienced Air Force fighter and bomber pilots. In terms of ongoing innovation in this field, the hunter–killer concept is very critical for future military conflicts. According to the director of the Strike Market Segment at the Northrop Grumman Integrated Systems in El Segundo, California, a true hunter–killer needs to be more than a weaponized ISR platform if the UAV has to perform the specialized tasks required by commandos. In summary, the hunter–killer concept requires to support the entire *kill* chain rather just the ISR mission [3].

If the battlefield commander is sending the hunter–killer UAV deep into a threat environment, he has to make sure before the assigned mission that the aircraft is fully equipped with the latest sensors and weapons and will not run out of bullets until the mission is completed. Note that putting more weapons in the UAV can adversely affect its endurance capability. Studies performed by the author reveal that there is a balance between the endurance and the sensor payload and the platform aerodynamic if stored externally. This will impact not only the endurance capability of the vehicle but the ability to transmit in battle-space, which is the most critical ability of how long will it take the UAV to reach a scene of activity. In order to provide precision strike capabilities in the global and the battlefield commander interest in reducing the *kill* chain consisting of find, fix, track, target, engage, and access with all of the information exchanges between the pilot and battlefield commander.

The MQ-9 is a hunter–killer UAV for USAF, but the Warrior is the hunter–killer platform optimized for a combination of ISR and attack missions. From the ground up, attack is the primary mission of the Warrior aircraft, according to U.S. Army commanders. However, the ISR capability can be used subject to commanders' discretion.

Frontline military experts consider the USAF's MQ-9 UAV and US Army's Warrior as precision weapon systems, not for broad-area attack activity. Since these UAVs carry *four* Hellfire missiles, you are therefore limited to four target attacks. Furthermore, there is a tradeoff between ISR and weaponization. As a matter of fact, the Warrior UAV is designed to carry a combination of EO/IR sensors, laser designator, SAR capable of see-through clouds, smoke, and fog, communication relay, and weapons specially designed and developed for this UAV. Essentially, the Warrior is a multimission platform with particular emphasis on the "kill-side" mission. Some battlefield experts debate whether to consider the hunter–killer a single platform or two distinct platforms—one for hunting the target and the other for killing the target. Regardless, one has to consider the cost involved and the operational complexity for the ground station crew. Operating a hunter–killer platform requires the ground station crew to be extremely alert at all times, which could make the crew mentally and physically tired.

Besides ground station crew operational tiredness, platform affordability is the most critical issue. If you want to deploy a single platform, you will need a larger vehicle as well as a large number of sensors leading to a very expensive system. Furthermore, USAF disagrees with a single-purpose killer, because this concept ignores the importance of the hunter mission. Studies must be undertaken whether it will be a cost-effective approach by keeping the fully loaded UAV for long endurance. In brief, one must give serious consideration for the deployment of a hunter–killer platform, which will not run out of both fuel and weapons. Preliminary studies performed by the author seem to indicate that there may be applications where a nonweaponized platform can be considered as a killer such as the U.S. Army's Apache helicopter. The studies further indicate that implementation of a laser designator on a killer platform will enable the pilot to detect, track, and kill. This approach will allow the killer platform to execute both hunter and killer missions with minimum cost and complexity.

Teaming and intelligent systems experts believe that there are so many different ways to develop the hunter–killer concept, and their variants will be coming in the near future. The U.S. Army Training and Doctrine Command has evaluated the concept of multiple platforms working together as a hunter–killer team in unmanned autonomous collaboration operations as demonstrated at the army base.

Autonomy of Hunter–Killer Platforms (MQ-9)

Military UAV experts believe that the potential autonomy of hunter–killer UAVs is not only impossible but also very controversial and complex. Program managers believe that regardless of the platform autonomy, keeping a man in the loop for making the "decision to strike" and "pulling the trigger" will remain unquestionably a mandatory part of any UAV or platform. These decisions require well-experienced operators (pilot and ground station operator). Any relaxation in the experience or training requirement of the operators could lead to critical mission failure.

Handling of multiple jobs can be most tiresome in a complex battlefield environment. For example, it will be too much of workload for the Apache pilot sitting alone in the front seat and controlling the helicopter and managing the sensors in a tense battlefield environment. Under such circumstances, teamwork or some sort of autonomy is essential for managing the multidimensional battlefield job. Availability of sophisticated decision-making software to handle minute-to-minute tasks or operation of a single UAV platform will increase the efficiency and management of the confidence level of the pilot.

To visualize the potential benefits of a single operator giving support to various unmanned platforms involving air and ground missions, a quasi-autonomous concept should be evaluated. This concept may not satisfy the requirements of hunter–killer missions. Another concept that may or may not fall into the hunter–killer category is the UCAV, sometimes seen as a long-term replacement for a manned fighter aircraft, but the USAF has rejected this concept. However, some tactical experts think that a UCAV aircraft can do as much as a hunter–killer MQ-9 can accomplish. In summary, it can be stated that a UCAV can provide the war fighter the essential performance capabilities in terms of efficiency, effectiveness, and tactical advantages close to those expected from the hunter–killer aircraft. Some military experts believe that a UCAV will perform better as a killer but can be designed for a long kill operation, where a lot of hunting is required.

It is important to mention that USAF uses active duty pilots who are well experienced as operators, while the U.S. Army has a tendency to seek high levels of platform autonomy, which allows the operators to handle more UCAVs with less direct input efforts. In summary, there is a gray area as far as the level of autonomy is concerned.

Role of Micro Air Vehicles

Technical articles published in *Defense Electronics* and *Military and Aerospace Electronics* journals seem to reveal that micro air vehicles (MAVs) are taking over some of the more dangerous missions for battlefield missions for the U.S. military. As a matter of fact, today, a number of small commercial air vehicles are operating without official permits or licenses. Note that MAVs can fall into three tier categories, namely, Tier-1, Tier-2, and Tier-3. These tier categories have been already described earlier in this chapter. Important characteristic features of these MAVs are briefly highlighted in Table 2.1.

Technical Specifications for Tier-1, Tier-2, and Tier-3 MAVs

The unmanned MAV technology has significantly matured during the past decade or so into a fully controllable and reusable combat aircraft. MAVs have proliferated into miniaturized forms and can be deployed in battlefield to undertake surveillance and reconnaissance missions without being seen or shot down. Massive deployment of MAVs in military applications is due to improvements in sensor and material technologies. Studies performed by the author seem to indicate that Tier-3 MAVs can be equipped with solid-state lasers, which can illuminate the target and can be destroyed by laser-guided Hellfire missile

Table 2.1 Outstanding Characteristics of Micro Air Vehicles

FEATURES	TIER-1	TIER-2	TIER-3
Manufacturer	Lockheed Martin AeroVironment	AAI Corp.	AAI Corp.
Altitude	Low altitude	Medium altitude	High altitude
Endurance	Short duration	Long duration	Long endurance
Weight	16 oz (max)	18 oz (approx.)	4.2 lb (max)
Mission	Surveillance	Surveillance	Reconnaissance
	Air targeting	Reconnaissance	Target search/track

Table 2.2 Technical Performance Specifications of Specific MAVs

MAV SPECIFICATIONS	WASP III	RAVEN RQ-11 B	PUMA AE
Weight (lb)	1.0 (max)	4.2	13.0
Wing span (ft)	2.375	4.5	9.2
Fuselage length (ft)	1.245	3.0	4.6
Typical speed (mph)	25–40	20–50	23–52
Operating range (mi)	3.0	6.0	9.0
Endurance (h)	0.75	1.0–1.5	2.0
		1.33–1.83 (with RB)	
Operating altitude (ft)	50–1000	100–500	300–1000
		14,000 (optional)	

Source: Editor of the journal, *Mach. Des.*, 46, 2009.
Note: The symbol RB stands for rechargeable battery. Now the author wishes to summarize the following performance capabilities of MAVs designed and developed by the AeroVironment company.

launched from other platforms. MAVs are best suited for reconnaissance or sustained surveillance. A MAV can be carried around in a backpack along with a ground-station controller and can be easily launched by hand. MAVs such as Wasp, Raven, and Puma have been manufactured by AeroVironment located in California. With their small size, ultralow RCR, and limited operating range, MAVs are very cost-effective in undertaking reconnaissance or sustained surveillance of targets in urban areas, over small military units, and in battlefield zones without being seen. The MAV Raven RQ-11 B has demonstrated range and endurance capability to provide military intelligence data in a variety of military missions such as target detection and tracking, force protection, convoy security, battle damage assessment, and tactical updates during conflicts in urban areas. Technical performance specifications of Wasp III, Raven RQ-11 B, and Puma AE are summarized in Table 2.2.

Wasp III MAV

The Wasp III and others have been designed and developed by AeroVironment of California. Note that Wasp III can be carried in a backpack and can be launched by hand. The Wasp III carries all the necessary sensors, including an altimeter, compact GPS-based navigation system, magnetometer, and nano-accelerometers. The most critical onboard component is the payload consisting of a pod with forward-looking or side-looking high-resolution EO color camera and IR camera. These cameras have limited tilt and zoom functions but have enough digital stabilization capability to keep the target in sight as the MAV flies in a nearby circular or in loiter patterns as it checks out the target position. Once the MAV is airborne, the vehicle follows a predetermined flight profile with a specified path and altitude. The operator can change the flight profile, if he wants, using a portable, ruggedized, handheld ground station to make its circle as the operator checks out the target. It is extremely important to mention that the portable ground station and the UAV remain in two-way secured communication, which means the MAV must remain within the LOS of the portable GCS as it beams back video signals. Generally, the UAV flies in an assigned altitude to a specified GPS waypoint and lands unless the operator takes over the MAV flight.

When the MAV needs to land, the operator specifies a landing spot, and the vehicle itself establishes a proper glide path and landing approach. As soon as the Wasp III calculates its distance from the ground and the right landing spot, it cuts the power of the engine, glides down, and slides across the landing area. It is important to know that the vehicle has no landing gear or wheels. According to published articles, the company is developing a version that can land on water. This requires that the aircraft body be waterproof and carry minimum additional weight that will not degrade its operating range or endurance capability.

Raven RQ-11 B MAV

This particular MAV has a wing span equal to twice that of Wasp III and a weight more than four times that of Wasp III. The vehicle uses a "pusher" propulsion unit mounted in the back of the main fuselage, as illustrated in Figure 2.8, because the customer wanted to have payload consisting of forward-looking and side-looking high-resolution color cameras and IR imagers. The payload is mounted on the vehicle so that the camera can look down the aircraft's flight path. Each payload item weighs about 6.5 oz, and the total vehicle weight including the sensors is close to 4.2 lb. The propulsion system for Raven RQ-11 B is much larger than that used for the Wasp III aircraft. Like most

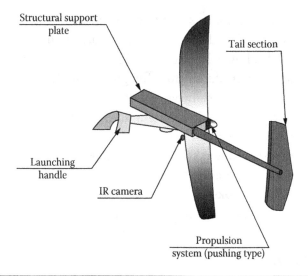

Structural support plate

Tail section

Launching handle

IR camera

Propulsion system (pushing type)

Figure 2.8   Structural details of a micro-UAV.

MAVs, the Raven flies according to visual flight rules (VFR) weather, which forbids it to fly in low clouds, high winds, heavy winds, or icing conditions. Because of its size, the MAV cannot be carried in a backpack and is controlled by the portable GCS operator or soldier. Note that the portable control system, including its hardware and software, has become a standard for military MAVs. Technical specifications and physical parameters of Raven are shown in Table 2.2. This MAV is widely used by the military for surveillance missions for locating ground targets and military building structures.

Puma AE MAV

The Puma AE is regarded as an all-weather MAV and has a larger wingspan, which offers large operating range and longer endurance capability to carry out most armed missions, including search, rescue, and drug interdiction functions. The Puma AE has been designed from the very beginning to be waterproof to meet the water-landing requirements without damaging the aircraft. If the Puma AE is filled with mud, dirt, or sand, it can be hosed off and then be ready to fly. Due to its large size and higher flight speeds, the Puma AE can fly under abnormal weather conditions, including heavy rain and heavy winds. Technical specifications and physical parameters of Puma AE are summarized in Table 2.2. Structural details of this MAV can be seen in Figure 2.8. This MAV is equipped with a mechanically gimballed payload, which includes a high-resolution EO camera, IR camera, and IR laser illuminator. The gimbal mechanism allows the operator to operate the cameras or laser illuminator through a 360° sector and tilt it from looking straight down to looking at the horizon position. It is reminded that the illuminator is not a target designator. It acts like a laser pointer, and it is boresighted with the high-resolution EO and IR cameras. It also lets the Puma MAV to put a spot on the actual target it looks at. Under these circumstances, the illuminator can be treated as a laser pointer and not a missile illuminator.

RQ-16 A T-Hawk

This is the latest MAV designed and developed by Honeywell Aerospace Company to meet U.S. Army and Navy requirements. It has neither a sleek design nor a streamlined form. Its principal advantages are vertical takeoff and landing and suitability for landing on ground or deck of a ship or aircraft carrier. Furthermore, it is ideal for landing in deserts, jungles, and on uneven roads. Its technical specifications and physical parameters are briefly described in Table 2.3.

This MAV runs on regular gasoline, a volatile fuel, and the military authorities would prefer its soldiers not to carry it into combat zones for personal security reasons. This miniaturized MAV deploys

Table 2.3 Technical Specifications and Physical Parameters of RQ-16 A T-Hawk

SPECIFICATIONS AND PARAMETERS	
Weight (lb)	17
Height (in.)	23 (max)
Diameter (in.)	13
Power plant	56 cc bower twin-piston engine
HP rating	4
Air speed (mph)	46
Operating altitude (ft)	10,000
Rate of climb (ft/s)	25
Guidance technique	GPS
Endurance capability	50 min or 38 mi

Source: Editor of the journal, *Mach. Des.*, 46, 2009.

a modified 4-HP two-stroke gas engine that runs a fixed-pitch ducted fan.

The spinning fan sends air through a tubular duct to generate the thrust force. Since the fan is enclosed in the duct, there is no danger to the soldiers while operating this MAV. In addition, the fan is protected from hitting any person or object while in the flight or landing. Note that the moving air creates a lift force as it passes over the airfoil lip of the duct. It appears that the duct design is extremely sound aerodynamically. Furthermore, the movable flaps at the bottom of the duct direct the thrust force for steering and stabilization functions. Note that the ascending or descending of the MAV is accomplished by increasing or decreasing the engine rpm. But the T-Hawk hovers at around 7000 rpm under normal wind conditions. The military is working on a 4 HP diesel engine or heavy-fuel engine, which will be safer as well as in compliance with the army's *one-fuel* initiative. Furthermore, a diesel-driven engine will be more cost-effective and extremely safe in handling the MAV in combat zones. On the other hand, the gasoline-driven engine is relatively quiet with the noise level not exceeding 60 dBA at an operating altitude of 400 ft. This level of noise is inaudible in most urban environments and will not bother anyone in the urban zones.

According to MAV designers, the MAV has not been flown by a pilot. However, the soldiers construct a flight path with up to 100 waypoints using a portable, rugged GCS. Note that the GCS maintains a two-way communication with the T-Hawk MAV for up to 5 mi using common military UAV frequencies. The MAV can store up to 10 flight paths. The GCS can store several hover, surveillance, and landing maneuvers that can be implemented any time. Also, the this vehicle has a capacity to download video signals at 1700 MHz and store up to 10 min of high-resolution images onboard the aircraft.

The payload of the T-Hawk vehicle includes an EO sensor and an IR sensor, and both are pointing forward and canted-forward. This type of sensor pointing allows the soldiers in the field to switch from one sensor to another in less than 15 s in the field. This vehicle is capable of detecting a man-size target at 820 ft during the day and at 320 ft during the night time. Note that providing zoom capability and vibration isolation features will lead to significant improvement in target detection and tracking.

Based on the aerodynamic features of the vehicle, it can be stated that this particular MAV aircraft can fly during day or night through rainfall not exceeding 0.5 in./h, a 23 mph wind, saltwater spray, fog, sand, or dust. The T-Hawk can take off or land in 17 mph wind environments. It can be carried in a backpack and can be launched in less than 5 min.

Small Tactical Munitions, Miniaturized Electronics, and Latest Component Technology for Future MAVs

Miniaturized EO sensors, miniature electronic components, and small tactical munitions must be available for the next generation of MAVs. Existing Hellfire missiles must be redesigned with emphasis on weight, size, and kill probability. Efficient and miniaturized electronic circuits for communication equipment should be given serious consideration. Gyroscopic and accelerometer designs must deploy MEMS-technology- and nanotechnology-based materials as much as possible to achieve significant reduction in weight and size of the components. It should be noted that reduction in weight and size and improvement in electronic components are of critical importance for the next generation of MAVs. Electronic circuit and component efficiencies can be improved using MEMS and nanotechnology materials. Reliability and mechanical integrity of structural components can be significantly improved using appropriate rare earth materials such as mu-metal.

A comprehensive review of a published technical article [7] by the author seems to indicate the deployment of additive manufacturing technologies in the design and development of small tactical munitions and critical components for MAVs. The newest tactical munition developed is known as Pyros. According to the Pyros designers, the munition developed is a very light, precise, and serious weapon. This weapon used multiple additive manufacturing technologies during prototyping and into the final production unit. The additive manufacturing technologies included fused deposition modeling (FDM) and selective laser sintering (SLS). Both these technologies offer high-quality materials highly resistant to adverse chemical effects and thermal environments. The FDM technology works via a heated nozzle, which extrudes material layer by layer, while the SLS technology works via a bed of powdered nylon and a carbon dioxide (CO_2) laser emitting at 10.2 μm wavelength, which sinters the material layer by layer. Note both these processes grow parts from the ground up, which provides part complexity that subtracting or eliminating technologies like machining find extremely difficult to emulate. Machining experts believe that additive manufacturing technology is regarded as an effective solution to the inhibitions of machining processing. Additive manufacturing technology provides better tolerances than machining. The drilling and milling of computerized CNC machining centers is very complex and costly. For complex parts, Pyros technology is quicker and cheaper especially for small tactical munitions like Pyros or even standard missiles such as Hellfire. Use of additive manufacturing technology provides consistent, tight tolerances.

Design engineers of Pyros believe that additive manufacturing technology helps in reducing the manual labor cost by integrating features directly into the part geometry such as critical attachment features and microfittings, mounting brackets, and control surfaces, which is otherwise not possible with minimum time and labor cost. Furthermore, this technology allows the manufacturing team to consolidate multiple features in one part of the tactical munition and provides them over incremental changes in the control surfaces and required tolerances.

Some munitions or weapons require fins for aerodynamic control and to maintain specific speed. Note that these fins help in steering the tactical weapon toward its target via two frames of reference that includes the GPS and a semiactive laser seeker or illuminator. It is important to mention that "with three-dimensional coordinates for its GPS, the Pyros weapon knows exactly where it is, thereby allowing the MAV operator to direct the weapon within 3 m of where the operator wants to be." Pyros designers claim that in dealing with moving ground targets or targets located within buildings, Pyros is equipped with a semiactive laser guidance system with an accuracy of 1 m. When the laser illuminates the target, the optical energy reflected off the target is used by the seeker to guide the weapon to the target. The needed direction and target information are processed simultaneously by the high-speed computer, and making accurate fin movement compatible with GPS and laser information is critical to the mission success.

Weapon manufacturing experts believe that application of the additive technology concept on different components and iterations of the Pyros weapon will provide a solid proof of benefits associated with this additive technology. Manufacturing engineers feel that sometimes rework may be required on critical components if the quality control specifications are not met or complied with. But that occurs very rarely. This problem was observed in the case of guiding fins on the weapon body. Guiding fins are imperatives in guiding the Pyros weapon to the target. Sometimes, it will be necessary to play with the control surfaces of the guiding fins till all specifications are satisfied by the quality control department.

Solid Concepts, Inc. is the manufacturing agent for the Pyros weapon. The company claims that additive technology plays an important role in weight reduction of small UAVs, which have payloads ranging from 5 to 100 lb. They have also demonstrated weight reduction capability for manned attack armed surveillance platforms.

Since most of the weight for Pyros comes from the warhead, other component weights must be subtracted to maintain the specified weight of the system. Material science indicates that the material composition of nylon, which is used in conjunction with SLS, yields components that are very light but still strong and extremely resistant to severe environments while incorporating more sophisticated features than conventional machining could hardly achieve in a single manufacturing operation.

Unmanned Ground Vehicles

For a couple of decades, defense scientists and engineers have been deeply involved in research, design, and development of UAVs, unmanned ground vehicles (UGVs), and UUWVs. Most of the unmanned vehicles are best suited for military applications as compared to civilian applications. The author feels that the U.S. Army will get the maximum benefits from such a research program, particularly in military conflicts.

A Pittsburgh, PA, company has been awarded additional funding for a contract by the U.S. Department of Defense to commercialize its high-speed inspection robot known as FrontRunner UGV. Note that the FrontRunner contract was awarded under the Small Business

Innovation Research (SBIR) program, but the additional funding was awarded through the Robotics Technology Consortium as a Phase Three SBIR contract.

The principal objective of this program was to provide the U.S. Army UGVs and UAVs that should truly be integrated and can be simultaneously controlled from a single operator from one GCS, providing optimum situational awareness and significantly improved mission effectiveness.

The principal goal was to demonstrate a true unmanned system, teaming capabilities and operational functions. Essentially, what it means is that the information gathered by a UAV, such as a potential roadside threat, can be used by the UGV to guide a conveyor to a safe location. This type of teaming capability will be found most suitable to move the conveyor to a safe location with minimum time and without any injury to soldiers in a hostile territory.

A sophisticated software program with joint architecture may be required with focus on integrating the ForeRunner UGV with the Insitu Common Open Mission Management Command and Control (ICOMC2) GCS. This control system will simultaneously control the FrontRunner and Insitu's unmanned aerial system (UAS). Note that the Insitu is a wholly owned subsidiary of the Boeing Aircraft Company [7]. Army program managers believe that in using SBIR III contract and in collaboration with Insitu, they are able to validate and demonstrate the benefits of a truly interoperable system.

Role of Unmanned Combat Aerial Vehicle in Counterterrorism

A comprehensive review of published papers on UCAVs [8] indicates that UCAVs have played a critical role in hostile lands starting from 2001. Particularly, the MQ-1 Predator aircraft armed with Hellfire laser-guided missiles are increasingly used in undertaking counterterrorism activities from the bases in Pakistan and Uzbekistan. These MQ-1 missions were aimed to eliminate the terrorist leaders and high-profile individuals inside Afghanistan's hostile regions. Some newspaper reports mentioned such attacks taking place in Pakistan, Afghanistan, Yemen, and Somalia, where the presence of high concentration of terrorist gangs was suspected. The principal advantage of using UAVs in such activities rather than the manned aircraft is to avoid a diplomatic embarrassment if the UCAV is shot down and the pilot is captured. Furthermore, such counterterrorism activities take place in friendly countries and without the official permissions from the countries involved. A Predator based in a friendly Arab country was used to eliminate a suspected Al-Qaeda agent in Yemen in 2002. This was the first time an armed Predator was used outside of a theater of war such as Afghanistan.

U.S. military authorities claim that the armed Predator strikes have killed at least *nine* senior Al-Qaeda leaders and dozens of low-ranking operatives. These attacks have demonstrated the most serious disruption of Al-Qaeda since 2001, when the Twin Towers were destroyed by the aircraft hijacked by the Al-Qaeda Islamic radicals. Terrorism experts believe that these antiterrorism attacks on Al-Qaeda leaders have confused the various terrorist groups so much that they are turning violently on one another. The terrorist leaders sincerely think that most terrorist activists are either dead or afraid to continue the dangerous activities.

As of October 2009, the CIA claims that they have killed more than 50% of the 20 most wanted Al-Qaeda terrorist suspects in aggressive operations using advanced UCAVs. Furthermore, as of May 2010, the counterterrorism experts believe that the UCAV strikes in the Pakistani tribal regions had killed more than 500 hardcore, Islamic militants. The officials believed that no more than 5% of nearby civilians including family members traveling with the targets were killed. As a matter of fact, the UCAVs linger overhead after a strike, in some cases for hours to count dead bodies including those of terrorists and civilians, if any.

In February 2013, U.S. Senator Lindsey Graham stated that about 4756 terrorists including civilians have been killed during the UCAV attacks. The effectiveness of these strikes was somewhat decreased when Pakistani intelligence officials tipped off terrorists prior to launches. The Bush administration decided in August 2008 to abandon the practice of obtaining Pakistani government permission before launching missile strikes from the Predator UCAV platforms. It is interesting to state that in the next 6 months the CIA carried out at least 38 successful Predator-based strikes in the northwest region of Pakistan, compared to 10 Predator strikes in 2006 and 2007 combined.

It is extremely important to mention that the effectiveness of Predator strikes is strictly dependent on the type of ammunition deployed and the caliber as well as the confidence level of the operator or the GCS pilot. Furthermore, the killing probability of the weapon is also dependent on the missile deployed. A preliminary weapon survey undertaken by the author indicates that the relative size of the weapon and its killing probability will determine the overall strike performance of the UCAV platform. The preliminary survey also reveals that laser-guided Hellfire missiles weigh about 100 lb, and they are strictly designed to eliminate tanks and attack bunkers. Smaller weapons including the Raytheon Griffin missiles and small tactical Pyros munition are being developed as a less indiscriminate alternate weapon system. Design and development activities are underway on still smaller U.S. Navy–developed Spike missiles. The payload-limited Predator A can also be armed with six Griffin missiles or only two heavier Hellfire missiles by the Predator platforms. In summary, it can be stated that Hellfire missiles are ideal for large-size UAVs, Griffin missiles are suitable for MAVs, and Navy-developed Spike missiles are best suited for deployment in nano-UAVs. However, the UAV operators and tactical experts are better qualified for the selection of missiles and weapon loads. UCAV experts believe that a payload-limited platform such as Predator A can be armed with six Griffin missiles as opposed to only too many heavier weapons such as Hellfire missiles.

Qualifications and Practical Experience for UAV Operators

The author wishes to comment on the academic qualifications and practical experience for handling the UCAV and UAV platforms versus manned fighter aircraft or manned bombers. According to combat experts, practical experience and operator caliber requirements must be given serious considerations for UCAV or UAV platform operators. As far as selection, type of weapons, and deployment of missiles are concerned, a manned aircraft pilot has multiple choices, where an unmanned platform operator has to use the weapons or missiles allocated according to the UAV platform weight and endurance restrictions.

However, academic qualifications for the UAV operators are important in selection and deployment of weapons or missiles, which will destroy the hostile targets with minimum collateral damage to property and civilians. Note that it will be extremely difficult for the UAV operator sitting in a GCS located from afar to differentiate between the targets and the civilian people, particularly if the civilians are living very close to the hostile targets. It should be mentioned that the UAV operator will make sincere efforts to direct the weapons on the hostile targets and not on the civilian population.

According to army combat officers, UAS operators used to working alongside combat troops will be found most valuable in accomplishing the most difficult missions. In other words, mission success could be 100% when the UCAV operator works closely with the combat troops. Furthermore, the U.S. Army commanders have not restricted themselves to using officers with piloting experience to operate their UCAVs. The U.S. Army training and doctrine command capability manager for the UASs confirms that this approach extends throughout the army. Furthermore, everything from the smallest to the largest systems is flown by the enlisted soldiers or operators.

Published technical articles reveal that the USAF initially deployed experienced pilots or seasoned aviators to operate the UCAVs platforms. But whatever the reasons may be, the Air Force recently started using listed men and women who have spent only a few tens of hours in the cockpits to fly its UAV platforms. The army selects the soldiers who finished the basic training period and trains them as UAS operators or GCS operators. Regardless of service philosophy, the military planners are able to work out combat-related tactics, techniques, and procedures for soldiers to use small aerial vehicles or MAVs in combat regions.

Summary

This chapter describes various critical issues associated with UAVs, and UCAVs are described for military applications with emphasis on sensor requirements and operator or pilot capabilities. Potential UAVs are identified for conducting identification, reconnaissance, and surveillance missions and combat missions. Performance capabilities for GCS operators or pilots with major emphasis on caliber and confidence of the operator are described. Roles of portable GCS and sensor requirements are briefly mentioned. Operational features of EO, electromagnetic, and physical sensors are summarized for the benefits of readers. Ideal locations for the GCSs are discussed. Deployment of COTS components for GCS and noncombat platforms are discussed with emphasis on cost, reliability, and performance. Sensor requirements for next-generation GCSs are defined with particular emphasis on cost and reliability. The types of sensors, weapons, and missiles aboard UCAVs are mentioned to deal with high-value targets. Performance capabilities of combat-related UAVs operated by various countries, including the United States, the United Kingdom, France, China, Israel, Russia, and India, are briefly summarized with emphasis on speed, range, ceiling, endurance, and power plant. Hunter–killer capabilities of UCAVs such as MQ-1 Predator, MQ-9 Reaper, and Sparrow Hawk are summarized with emphasis on range, service ceiling, endurance, and sensors deployed. The level of autonomy for hunter–killer platforms is briefly discussed with emphasis on operator confidence level, "decision to strike," and "pulling the trigger." The physical dimensions and performance parameters of nano- and micro-UAVs are briefly described on weight, range, and endurance. Miniaturized EO sensors such as mini-SARs and uncooled forward-looking and side-looking IR cameras for applications in miniaturized UAVs are described with emphasis on weight, power consumption, and performance parameters such as pixel size and pitch. Technical sensor specifications for Tier-1, Tier-2, and Tier-3 MAVs are briefly summarized with particular emphasis on size, weight, range, endurance, and mission objective.

References

1. Wikipedia, Unmanned aerial vehicles for military applications, Wikimedia Foundation, Inc.
2. Y. Azoolai, Unmanned aerial vehicles shaping future warfare, *GLOBES*, October 24, 2011, pp. 9–13.
3. J.R. Wilson, Hunter-killer UAVs to swarm battlefields, *Military and Aerospace Electronics*, July 2007, pp. 2–9.
4. J. Mohale, Cockpits on ground, *Military and Aerospace Electronics*, July 2010, pp. 18–23.
5. Tech Briefs, Queuing model for supervisory control for unmanned autonomous vehicles, *Aerospace and Defense Technology*, Space and Naval Warfare Systems Center Pacific, San Diego, CA, May 2014, pp. 33–34.

6. Editor of the journal, Remarks on special aerospace and defense, *Machine Design*, 46–51, December 10, 2009.
7. Defense Technology Sector, Unmanned demonstrator aircraft for maritime surveillance, *Aerospace and Defense Technology*, May 2014, pp. 38–40.
8. J. Browne, UAVs leading forward ranks, *Defense Electronics*, 36–39, September/October 2010.

3

ELECTRO-OPTICAL, RADIO-FREQUENCY, AND ELECTRONIC COMPONENTS FOR UNMANNED AERIAL VEHICLES

Introduction

This chapter is dedicated to the electro-optical (EO), radio-frequency (RF), and electronic devices or components best suited for the current and next generations of unmanned aerial vehicles (UAVs) and unmanned combat aerial vehicles (UCAVs). Reliability, device performance, miniaturization, low power consumption, and minimum weight and size have been given serious consideration in identifying these components for UAVs. In addition, critical parameters such as extreme thermal environments, mechanical integrity, shocks, and vibration are considered in the component selection process, because these UAVs may be required to operate at altitudes ranging 5,000–50,000 ft in military missions. Fuel consumption is strictly dependent on the vehicle weight, weapon payload, endurance, flight dynamic conditions of the platform, and the weight of the EO and RF subsystems such as forward-looking infrared (FLIR) and IR high-resolution cameras, associated stabilizing gimbals, side-looking radar (SLR), illuminating IR laser system, reconnaissance receiver, surveillance receiver, pods for special sensors, laser-guided Hellfire missiles, and so on. Because of atmospheric turbulence and unstable flight conditions, the EO and RF sensors must be secured at the appropriate locations on the UAV platform using strong, reliable, miniaturized bolts, lock-nuts, and screws.

To retain mechanical integrity and the required RF and IR performance of the sensors aboard the UAV or UCAV platform, miniaturized RF coaxial connectors, edge launch RF connectors, bulkhead and panel adaptors, and customized cable connectors play critical roles. Each category of connectors, adaptors, and various types of compact DC power sources or compact power supplies with multiple output voltages and rechargeable batteries best suited for UAVs will be described with particular emphasis on system performance, reliability, weight and size of the EO and IR sensors, reconnaissance receivers, surveillance receivers, and high-resolution, compact SLRs capable of meeting the critical functions of the UAV combat mission such as target detection, identification, and tracking function. Besides reduced weight, size, and power consumption, EO/IR components and mechanical parts must meet stringent reliability, safety, and longevity requirements. Furthermore, all analog and digital devices, tactical data links, printed circuit antennas, missiles, computers, error-free algorithms, softwares, communication equipment, and a host of other UAV components must meet stringent performance, reliability, and structural integrity requirements under severe thermal and unstable aerodynamic environments. The next generation of UAVs and UCAVs is expected to be designed for longer endurance capabilities to minimize cost and to satisfy other military and border-related security requirements.

RF Components for UAV and UCAV Sensors

From here on, the author will attempt to focus briefly on unique performance characteristics and physical parameters of the sensors and subsystems. Types of RF connectors will be described with particular emphasis on the component's mechanical integrity, weight, size, and reliability needed in critical combat missions.

RF and Microwave Passive Components

RF and microwave components widely deployed for UAV microwave sensors will be discussed with emphasis on component weight, shape, and size, which impact the reliability, mechanical integrity, longevity, and endurance capability of the UAV platform.

Synthetic Aperture Radar, a Premium Sensor for UAVs

This sensor is best suited for achieving high-resolution images of targets and detection and identification of airborne and ground moving targets with high probability. Note that this system is capable of operating day or night under all weather conditions such as rain, snow, fog, and smoke. Compact printed circuit antenna elements of the synthetic aperture radar (SAR) [1] are fed by the solid-state transmitter that will significantly reduce the sensor weight and size, leading to improved reliability and endurance of the platform. Figures 3.1 and 3.2 show various types of RF connectors, cable connectors, adaptors, and bulkhead connectors, which can be used to install the transmitter package and antenna assembly with minimum size and weight. The micro-RF connectors shown in Figure 3.3 will ensure highest reliability, improved structural integrity, and significant reduction in sensor weight. It is interesting to point out that the U.S. Army has funded well-established microwave companies to develop most compact subsystems for deployment in UCAVs or UAVs such as NANO-SARs, low-noise reconnaissance and surveillance receivers, and digital RF memory (DRFM), which is a critical component of electronic warfare equipment with minimum weight and size. For the benefit of the readers, it is desirable to have the NANO-SAR sensors designed and developed particularly for UAV

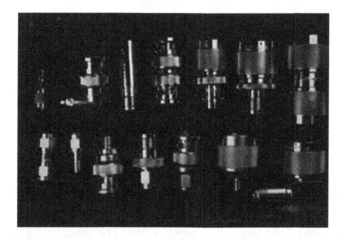

Figure 3.1   Precision radio-frequency connectors with high integrity.

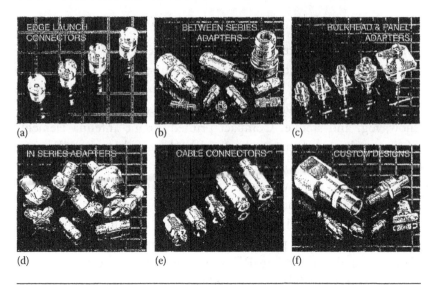

Figure 3.2 Precision connectors with high quality and reliability. (a) Edge launch connectors, (b) between-series adapters, (c) bulkhead and panel adapters, (d) in-series adapters, (e) cable connectors, and (f) custom design.

Figure 3.3 Miniaturized connectors with high accuracy, reliable performance, and impressive repeatability, using deep-drawn manufacturing technology.

platforms to conduct reconnaissance and surveillance missions with utmost precision. As far as the SAR operating performance principle is concerned, it is important to state that the sensor uses the relative motion between the UAV platform and the target. This provides long-term coherence signal variations, yielding fine spatial resolution needed for target recognition. In brief, the NANO-SAR provides target detection, recognition, and classification functions using real-time aerial images strictly due to deployment of the state-of-the-art RF passive components and miniaturized electronic parts for secured communication equipment.

NANO-SAR Performance Parameters Deployment of this sensor in a UCAV is tantamount to military victory in a life and death situation. This most compact system offers the following unique physical and performance parameters [1]:

- *Resolution*: 1, 1.65, 3, 4 ft
- *Maximum power consumption*: 25 W
- *Solid-state transmitter output power*: 1 W (min)
- *Antenna dimensions*: 5.5 × 3.5 × 2.0 in
- *Space occupied by miscellaneous item such as cables, turret, etc*: 85 in.³

Note that the SAR is integrated with Lisa ground station and Viper communication link to provide real-time monitoring of the performance of SAR and data link. SAR designers reveal that the sensor can be installed in an appropriate section of the fuselage, which could significantly improve the endurance capability of the platform.

RF Components for Reconnaissance and Surveillance Receivers

RF connectors for use in microwave reconnaissance and surveillance receivers could deploy edge launch connectors, custom-designed connectors, and precision coaxial connectors as shown in Figure 3.2. Studies performed on miniaturized parts best suited for electronic, communication, and aerospace applications seem to indicate that deep-drawn microparts shown in Figure 3.3 will provide the best overall performance capabilities. It is extremely important to mention that the precision deep-drawn process can provide enhanced production efficiencies and significant cost savings over traditional manufacturing methods. In brief, the deep-drawn microparts available from Braxton manufacturing company are best suited for low-noise reconnaissance and surveillance receivers, where frequency resolution, low FM noise level, and wide instantaneous bandwidth are principal specification requirements. These microparts are

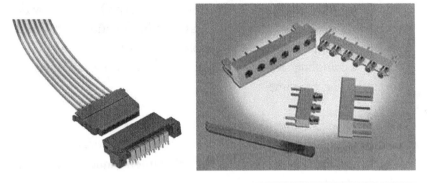

Figure 3.4 Multiport radio-frequency connectors for electronic and communication applications.

capable of providing high accuracy and repeatability with tolerances better than ±0.0038 mm (0.000150 in or 150 μin) and surface finish better than 120 μin. The studies further indicate that these precision parts are widely used for electronic, communication, and critical aerospace applications, where precision, accuracy, and reliability are considered as the highest priorities. Our research studies reveal that multiport RF connectors, as shown in Figure 3.4, could be found to be most appropriate for sophisticated electronic, tactical data link, and secured communication systems, where security, reliability, RF leakage, and connection integrity are the highest priorities.

Connectors and Cables for Tactical Data Link

Tactical data link is of critical importance to the ground control operators who are continuously monitoring the performance of the UAV or UCAV platforms. The data link provides RF data collected by SAR to a satellite. If the UAV is deeply involved in the reconnaissance and surveillance missions in conflict zones, secured communication will be absolutely necessary. The UCAV must be provided with a secured data link and precision RF leakage-free connectors are deployed to retain the highest integrity of the data collected by the SAR system.

Cables and connectors must have minimum insertion loss (IL) and low voltage standing wave ratio (VSWR) and must be free from sharp bend. The RF technician must make sure that the tactical data link is rigidly connected to the subsystem and free from RF leakage occurring from the cables or connectors. The performance tactical data link can be highlighted as follows:

- Minimum link margin (dB)
- Penetration from an external RF source through the RF connectors
- Frequency modulation due to vibration from a sensor or platform
- Data transmission or voice communication that must be done with minimum time
- Typical insertion loss (IL)
- Phase stability versus temperature
- Operating frequency upper limit
- Phase stability as a function of flexure

Data Security

Data security is virtually the principal requirement for all military systems. Note that compromising data in commercial transactions could mean financial loss, whereas it could mean mission failure or loss of the costly UCAV vehicle in a military operation. In summary, it is absolutely essential to protect and preserve the integrity of the military data and must not be compromised under any circumstances. Modern databases and data are synonymous with computers, softwares, and operating systems; therefore, they must be constantly checked for weaknesses. For specific details, refer to the DoD's Defense Security Services (DSS) that provide excellent guidance on data security. In addition, the latest white papers provide up-to-date information on threats or computer viruses and how to achieve the highest levels of data security.

Semiactive Passive Microwave Components for UAVs

In this section, quasi-active or semiactive RF components such as semiconductor-based limiters, ferrite-based limiters, and yttrium-iron-garnet (YIG)-tuned RF filters will be described with particular emphasis on weight and size. RF components will be described that are best suited for UCAV vehicles. Important characteristics and performance capabilities of such components will be briefly described with emphasis on reliability, safety, and endurance capability. Two distinct types of limiters, namely, semiconductor-based limiters and ferrite-based limiters, will be described in great detail with particular emphasis on IL, limiting threshold, protection level available and limiting dynamic range. As mentioned earlier, these limiters could provide the needed protection for the sensitive reconnaissance and surveillance receivers widely used by UCAVs.

Semiconductor-Based Limiters

The semiconductor-based limiter can be designed using varactor diodes, PIN diodes or step-recovery diodes. Cost, limiting threshold, DC bias requirement, and limiting dynamic range are strictly dependent on the type of semiconductor diode deployed. Limiters can be classified into two distinct categories, namely, soft limiters and hard limiters. Limiters designed with low threshold levels are characterized as soft limiters, and such limiters use semiconductor diodes. The limiters developed with high threshold levels are considered as hard limiters, and these limiters use miniaturized ferrite spheres. Soft limiters can be designed with a threshold level ranging from 0.01 to 10 mW. Soft limiters can be designed and developed using high-Q varactor diodes or PIN diodes or step-recovery diodes with appropriate DC bias circuitry. Typical performance parameters of a soft limiter using high-Q varactor diodes can be summarized as follows [2]:

- *Maximum input VSWR*: 1.2:1 over 10%–20% bandwidth at S-band and X-band frequencies
- *Typical IL*: 0.5 dB
- *Threshold level*: 0.001 mW (min)
- *Limiting range*: 30–50 dB (typical)

Ferrite RF Limiters

Diode limiters are mostly used to protect the sensitive receivers such as reconnaissance or surveillance or any sensitive microwave receiver, whereas the ferrite RF limiters are normally deployed for the protection of high-power radar duplexers. Essentially, the ferrite power limiters are used in radar signals to enhance the RF performance of the duplexer. In brief, the limiter provides improved receiver crystal-diode burnout protection during the transmission. Note that low-power ferrite limiters can be used to protect the receiver from jamming signals.

As mentioned earlier, ferromagnetic nonlinearity property is used to develop gyromagnetic coupling to generate the limiting action in a material. Note that the ferrite material is used with narrow linewidth, which is biased to enhance the limiting action. Ferrite limiters are mostly used in the protection of radar duplexers because of higher threshold requirement. Typical ferrite limiter performance specifications can be summarized as follows:

- *Ferrite material linewidth (max)*: 0.5 Oe
- *Limiting threshold range*: 100 mW–50 W
- *Peak input power range*: 1–30 kW (the latest limiter designed with input power as high as 100 kW)
- *Maximum IL*: 1 dB
- *Instantaneous bandwidth*: 15 (max) for S- to Ku-band frequency
- *Operating temperature range*: 0°C–85°C
- *Latest limiting dynamic range*: 50 dB (typical)

Note that nonlinear properties [3] are seriously affected at higher operating temperatures, which could affect the IL and dynamic range. Furthermore, excessive temperature could produce the instability if adequate cooling is not available to the ferrite limiters.

Yttrium-Iron-Garnet-Tunable Filters

The YIG-tunable filter plays a critical role in protecting the sensitive sensors such as radar and sensitive receivers aboard the UCAV. Fast tuning rate with minimum IL is the most desirable performance parameter for the YIG-tunable filter to protect the sensitive equipment aboard the UCAV. When a radar or a reconnaissance receiver aboard the UCAV is being jammed by the enemy, the YIG-tunable filters protect the UCAV sensitive sensors using high attenuation ranging from 40 to 60 dB at the jamming frequencies. These YIG filters are electronically tunable over the wide frequency spectrum. Fast tuning capability and high-stop attenuation capability are the principal design requirements of such filters. It is interesting to point out that the YIG-tunable filter is referred as the magnetically tunable (MT) filter. Its specific design aspects are shown in Figure 3.5.

Working Principle of a Magnetically Tunable Filter   An MT filter can be achieved using single-crystal YIG ferromagnetic resonators in the form of microspheres with a diameter ranging from 0.050 to 0.100 in. These spheres must have polished surfaces to achieve a high-quality value known as Q. These YIG spheres or disks act like high Q microwave cavities, which can be electronically tuned over wide microwave frequencies with minimum IL. However, low IL requires a very high-quality factor "Q" ranging from 5,000 to 10,000. It is extremely important to mention that YIG spheres must be located in the vicinity of a strong microwave magnetic field. Coupling from one YIG sphere to another can be changed by means of a loop or aperture. The resonant frequency can be changed by varying the strength of the externally applied magnetic field. YIG band-pass filters are available for various octave bands. This tunable filter can use spherical YIG microelements or YIG flat elements as illustrated in Figure 3.5. Performance parameters of the YIG filter can be summarized as follows:

- *Tuning range*: Octave or more
- *Instantaneous bandwidth*: 5–60 MHz
- *X-band 3 dB nominal bandwidth*: 30 MHz
- *Maximum IL*: 2 dB
- *Minimum off-resonance attenuation*: 50 dB or more
- *Maximum input VSWR*: 1.5:1 when tuned over the entire X-band region (8–12.4 GHz)

Solid-State Tunable Oscillators for UAV Applications   A tunable solid-state oscillator can be used as a low-power local oscillator for reconnaissance and surveillance receivers widely used by the UAV platforms. In case of a tunable phase-locked oscillator, the frequency change is made in discrete steps; therefore, phase lock can be accomplished at the discrete frequency. This particular technique offers the highest frequency stability and accuracy. Deployment of an operational amplifier provides the feedback loop with the desired roll-off characteristics. Design studies performed by the author in the past reveal

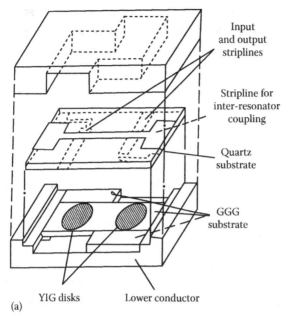

Input
and output
striplines

Stripline for
inter-resonator
coupling

Quartz
substrate

GGG
substrate

YIG disks   Lower conductor

(a)

RF performance parameters:

Maximum insertion loss = 2.20 dB

Stopband rejection = 50 dB

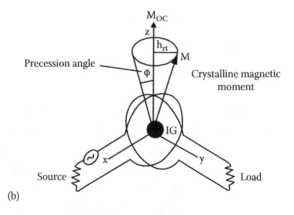

(b)

Figure 3.5 Magnetically tunable microwave band-pass filters using yttrium-iron-garnet disks or spheres. (a) Internal details of the filter's physical and radio-frequency (RF) parameters: RF tuning range, 14.1–15.7 GHz. (b) Interaction between applied magnetic field and RF sign.

that greater phase stability is obtained at lower oscillator frequencies, while the highest oscillator frequency is achievable through the phase lock technology. Remember that both the frequency stability and the phase stability are desirable for the UAV reconnaissance and surveillance receivers. In summary, solid-state transistor-based tunable oscillators are most ideal as local oscillators for possible application in the sensitive UAV receiver systems.

Reconnaissance and Surveillance Receivers

Studies performed by the author reveal that all receivers aboard UAVs and UCAVs for providing identification, reconnaissance, and communication functions should meet minimum noise and bandwidth requirements besides other performance specifications. Most low-noise receivers are protected by solid-state diode-based limiting devices with minimum IL. Research studies undertaken by the author seem to reveal that low-noise reconnaissance and surveillance receivers should meet the following performance parameters:

- Q maximum at room temperature
- Dynamic range
- Instantaneous bandwidth
- Tuning range
- Phase and amplitude instabilities
- Signal-to-noise ratio (S/N)
- Instantaneous bandwidth
- IF bandwidth

The research studies further indicate that receiver noise figure, dynamic range, and S/N ratio are the most critical performance parameters. The noise could be reduced by cryogenic cooling that will increase receiver cost, complexity, weight, and size; therefore, this option could not be justified for the receivers aboard UAV platforms. Research studies indicate that the receiver noise figure and dynamic range cannot be optimized simultaneously. The receiver dynamic range decreases with increasing IF bandwidth. For the benefits of the readers and design engineers, typical room-temperature noise figure and dynamic range for various microwave receivers are shown in Table 3.1.

Table 3.1 Typical Room Temperature Noise Figure and Dynamic Range for Various Microwave Receivers at 10 GHz Frequency

PARAMETERS FOR VARIOUS MICROWAVE RECEIVERS	NOISE FIGURE (dB)	DYNAMIC RANGE (dB)
Uncooled parametric amplifiers	3.2	37
Tunnel diode amplifiers	5.3	32
Transistor amplifiers	9.0	56
Balanced diode mixer amplifiers	8.5	65
GaAs MMIC amplifiers	4.5	40

The noise figures of GaAs monolithic microwave integrated circuit (MMIC) amplifiers are significantly lower compared to all other receivers as shown in Table 3.1, except uncooled amplifiers. The GaAs MMIC uses MESFET devices, and the MMIC amplifier noise figure is roughly 4.5 dB at room temperature [4]. The MMIC amplifier uses a MESFET with transistor length and width dimensions of 0.15 μm × 50 μm, approximately. The MMIC amplifier room temperature noise figure and gain are 4.5 and 30 dB, respectively, when operating over a frequency range of 8–18 GHz. MMIC amplifiers using high-electron-mobility transistor (HEMT) devices are capable of operating over multiple bands (2–21 GHz) with noise figure not exceeding 3 dB. Major advantages of GaAs MMIC amplifiers can be summarized as follows:

- Small weight and size
- Improved package reliability
- High mechanical integrity
- Higher gain greater than 55 dB
- Maximum room temperature noise figure (4.5 dB)
- Improved AM-to-PM amplifier performance (less than 0.5 deg/dB)
- Gain flatness over the entire band (±1 dB)
- Excellent amplitude and phase stabilities
- Amplitude tracking (±0.5 dB)
- Phase tracking better than ±4°
- Power-added efficiencies (greater than 60%)

Note that excellent amplitude tracking and peak power-added efficiencies are available over the narrowband but not exceeding 15% bandwidth.

Low-Noise MMIC Amplifiers

MMIC amplifiers use MESFETs and are best suited for applications where minimum weight and size are the principal requirements [5]. Note that MMIC amplifiers could be most ideal for UAV applications, provided they meet all performance specification requirements. In some cases, the cost of MMIC amplifiers may make them impractical for use. The cost of a microwave component or circuit strictly depends on five distinct factors, namely, yield, size, material, production volume, and automation. The production cost of a hybrid component is lower compared to that of an MMIC-based component. Studies undertaken by the author on hybrid and MMIC circuits or components indicate that testing for reliability, reproducibility, and performance verification requires much higher costs irrespective of the component technology. However, MMIC technology will be most attractive for a mass-scale production of microwave amplifiers for possible applications in current and future UCAV military operations. Studies undertaken by the author reveal that narrowband MMIC amplifiers are best suited for satellite and point-to-point communication systems, and their performance parameters are summarized in Table 3.2. The studies further reveal that broadband MMIC amplifiers are most suitable for the current and next generations of UAVs or UCAVs, and their performance characteristics are summarized in Table 3.3. Research studies performed seem to indicate that MMIC amplifiers using HEMTs have demonstrated excellent performance for electronic countermeasures (ECMs), electronic support measures (ESMs), reconnaissance, and surveillance missions to be carried out by the

Table 3.2 Typical Performance Parameters of Narrowband MMIC Amplifiers

AMPLIFIER PARAMETERS	UNIT 1	UNIT 2	UNIT 3
Frequency range (MHz)	3700–4200	10,700–11,700	14,000–14,500
Output power (W)	2	12.5	12.5
Stable gain (dB)	30	5.6	5.0
Power-added efficiency (%)	65	25	21
AM-to-PM amplifier performance (deg/dB)	0.25	0.50	0.75
Operating temperature (°C)	−55–100	−55–100	−55–100

Source: Jha, A.R., *Monolithic Microwave Integrated Circuit (MMIC): Technology and Design*, Artech House, Inc., Norwood, MA, 1989, pp. 10–25.

Table 3.3 Performance Parameters of Wideband Monolithic Microwave Integrated Circuit Amplifiers for Unmanned Aerial Vehicle–Based Reconnaissance, Surveillance, Electronic Countermeasure, and Electronic Support Measure Missions

PARAMETERS	UNIT 1	UNIT 2	UNIT 3	UNIT 4	UNIT 5
Frequency range (GHz)	0.1–12	2–18	6–18	2–21	3–40
Output power (W)	10	13	30	18	15
Gain (dB)	16	11	30	12	6
Noise figure (dB)	4.5	7.5	8.5	3.0	4
Monolithic device used	MESFET	MESFET	MESFET	HEMT	HEMT
Application best suited for	ESM	ECM	ECM	ECM	ECM
Operating temperature (°C)	−55–125	−55–125	−55–125	−55–100	−55–100

Table 3.4 Power Added Efficiencies for Various Discrete Devices as a Function Frequency (%)

FREQUENCY (GHz)	DEVICE TYPE		
	MESFET	HEMT	p-HEMT
18	38	44	59
35	30	35	42
60	12	16	28
95		5	18

current and next generations of UCAVs. Performance capabilities and limitations of wideband MMIC amplifiers using HEMTs are summarized in Table 3.4.

Performance Parameters of MMIC Amplifiers for
Deployment in the Next Generation of UCAVs

It will be interesting to know the critical performance parameters and their limitations of GaAs MMIC amplifiers operating over narrow and wide instantaneous bandwidths and using MESFET and HEMT devices. Critical performance characteristics of various MMIC amplifiers are summarized in Tables 3.2 and 3.3.

Typical performance parameters of wideband MMIC amplifiers for the current and next generations of UCAVs are shown in Table 3.3. As mentioned previously, these wideband MMIC amplifiers are most suitable for reconnaissance, surveillance, ECM, and ESM mission applications.

Reliability and Structural Integrity of the
Transistors Used in MMIC Amplifiers

Two types of transistor devices are used in MMIC amplifiers, namely, MESFET devices and HEMT devices. Both reliability and structural integrity are strictly dependent on metallization, dielectric substrate, semiconductor material, and packaging structure. The appropriate thickness of a metal layer should be used to achieve higher reliability. Note that integration of active and passive elements on the same chip offers increased performance and reliability with minimum cost and circuit complexity. MMIC design engineers claim that its reliability is strictly dependent on the reliability of active and passive elements, such as field effect transistors (FETs), printed circuit inductors and capacitors, interconnect lines, crossovers, via holes, and other components used in the design and development of MMIC assembly. In case of MMIC amplifiers, main reliability concerns are due to the following sources:

- Ohmic contact metallization
- MESFET channel conduction integrity
- Electromigration of gate, first-level interconnect, and air-bridge metallization
- Thermal diffusion in the substrate
- Burnt-out problem at higher input pulses
- Mechanical failures in package sealing
- Failures due to adverse channel temperatures

In summary, failure mechanisms are mainly due to metallization, dielectric, semiconductor material, and the discrete devices such as diodes or transistors. Solid-state scientists believe that greater integration increases the package density, which introduces different failure mechanisms such as coupling between the closely spaced components and interconnects, electromigration in narrow interconnect lines, and increased power dissipation leading to device failure.

Power performance of MMICs is strictly dependent on the types of discrete devices used in the MMIC assembly. There are three types of transistors that can be used in the design of an MMIC amplifier such as MESFET, HEMT, or pseudomorphic HEMT (p-HEMT).

Table 3.5 Performance Parameters of MESFET, HEMT, and p-HEMT Transistors

TRANSISTOR TYPE	FREQUENCY (GHz)	MESFET	HEMT	p-HEMT
Gate length (μm)		0.25	0.25	0.25
Cutoff frequency (f_{co})	Gigahertz	108	170	230
Frequency (GHz)	18	1.3	0.91	0.85
	35	2.40	1.75	1.55
Noise figure (dB)	60	3.55	2.54	2.35
	94	4.90	3.50	3.30
Frequency (GHz)	18	15.56	19.50	22.13
	35	9.79	13.73	16.35
Available gain (dB)	60	5.11	9.05	11.67
	94	1.21	5.15	7.77
	143		1.50	4.13
Power-added efficiency at 60 GHz (%)		12	16	28 (min)
				36 (max)

Power-added efficiency is the most critical performance parameter of a power amplifier, and the efficiency values of these devices at microwave and millimeter-wave frequencies are summarized in Table 3.4.

Both the power output and power-added efficiency of the three devices are function of device type, gate dimensions, and cutoff frequency (F_{co}). The cutoff frequency of the device is strictly a function of saturation electron velocity.

Both HEMT and p-HEMT transistors are best suited for millimeter-wave, wideband, low-noise reconnaissance, and surveillance amplifiers, which are widely used by UAV and UCAV platforms. Note that the noise of these amplifiers, maximum available gain, and power-added efficiency are strictly functions of the gate length and cutoff frequency of the transistor. Calculated magnitudes of these parameters at millimeter-wave frequencies are shown in Table 3.5.

Electro-Optical Sensors for UAVs

The subsection will deal with EO sensors such as lasers, radars, FLIR sensors, high-resolution IR cameras, radar-guided missiles, laser-guided missiles, and focal planar detectors for undertaking specific missions by UCAVs. Performance capabilities and critical role of each sensor in deep penetration of the UAV and UCAV missions in the enemy territory will be briefly discussed with emphasis on reliability and safety.

Lasers and Their Critical Roles in UAVs

Laser is defined as light amplification by stimulated emission of radiation. There are various types of lasers available such as gas or solid state lasers. The output wavelength could be in the microwave, IR, visible, ultraviolet, or x-ray region of the spectrum. Lasers can be classified as chemical lasers, gas lasers, semiconductor lasers (also known as solid-state lasers), and rare earth-based lasers, namely, lithium-yttrium-fluoride (LYF) lasers, lamp lasers, and so on. The author will identify and briefly describe laser systems with minimum weight and size and optimum laser performance. Chemical lasers are ruled out because of excessive weight and size and chemical storage problems and therefore are not suitable for UAV applications. Regardless of the laser type, all EO signals experience absorption, scattering, reflection, and diffusion while travelling through the atmosphere. Semiconductor lasers and rare earth crystal-based lasers could be more ideal for UAV applications, if their efficiencies and power output levels meet the laser system requirements at IR wavelengths. Note that high-power laser beams are seriously affected during transmission in the various regions of the atmosphere. Space scientists and theoretical physicists believe that high-power gaseous lasers and chemical lasers (hydrogen fluoride [HF] and deuterium fluoride [DF] lasers) are best suited for space applications, if safety problems are adequately addressed.

Transmission of IR radiation in the atmosphere is very complex due to the dependence on scattering and absorption effects. Even in clear troposphere, IR radiation is attenuated due to the absorption and scattering by the atmospheric gases and aerosol particles. Note that the atmospheric absorption is primarily caused by the presence of molecular absorption bonds and is strictly a function of IR radiation wavelength. The amount of attenuation introduced by the atmospheric absorption and scattering is an important factor in the design and selection of appropriate lasers best suited for UAV applications. Atmospheric attenuation reduces laser system performance particularly at lower altitudes. But in the case of target tracking and acquisition missions, the molecules can seriously degrade the laser system performance, leading to mission failure. Even in clear atmosphere with no suspended particles such as dust, smoke, fog, or rain, transmission of laser signals is affected. In summary, laser beam transmission in the atmosphere is strictly dependent on IR laser wavelength, propagation path, operating altitude, seasonal conditions, and variations in atmospheric temperature and air density. At high-power levels, laser beam propagation is subjected to more prominent effects turbulence-induced beam spreading and wandering. Various types of lasers can be classified as follows [6]:

- Chemical lasers such as DF and HF lasers
- Gas lasers such as carbon dioxide (CO_2) lasers
- Semiconductor lasers such as GaAs injection lasers
- Rare earth crystal-based lasers such as LYF lasers
- Diode-pumped solid-state (DPSS) lasers
- Dye lasers
- Flash-pumped solid-state lasers
- Excimer lasers
- Chemical oxygen iodine lasers (COILs)

Research studies undertaken by the author on potential lasers for UAV applications indicate that laser requirements for target illumination are different from the laser needed for target detection, tracking, and identification purposes. A comprehensive review of the latest technical research papers may lead to identifying the specified laser type that could satisfy both requirements.

Laser Seeker for UAV Applications

A laser seeker is a passive EO sensor capable of detecting laser energy reflected from a target, which is illuminated by a target designator. This sensor can be mounted under the UAB belly or in the nose section. When a UAV equipped is with both the laser seeker and the laser illuminator, it can provide effective close support in target identification, nighttime interdiction, precision weapon delivery, and accurate attack capability under fair weather conditions. Note that laser-guided weapons yield low circular error probability (CEP) not exceeding 15–20 ft. The IR seeker is the most critical element in a laser-guided bomb (LGB) or IR missile. Note that the laser seeker must operate at an appropriate IR wavelength ranging from 3.2 to 4.3 μm for optimum performance. This particular IR window is preferred to minimize false targets from clouds and terrestrial objects. The laser seeker must deploy the following state-of-the-art components and advanced features:

- Large field of view (FOV)
- Small instantaneous FOV (IFOV)
- High-resolution focal planar array (FPA)
- *Number of pixels available*: 256 × 256 or 640 × 256
- Laser seeker performance being dependent on visibility and target reflectivity

Laser Illuminator

Target detection and identification and the performance of LGBs are strictly dependent on the performance of the laser illuminator. As a matter of fact, the sensitivity of the laser illuminator is dependent on the optical reflections from the target, atmospheric conditions, distance to the target, and the laser illuminator CW output power, which typically varies from 50 to 300 W. The reflections from the target depend on the target size, reflectance, and emissivity of the target surface material. Operating range capability is dependent on atmospheric conditions, such as fog, cloud, or smog, and scattering from aerosols and turbulent conditions. The laser illuminator or designator can be designed to operate in CW mode or in a high-duty pulsed mode depending on the cost, weight, size, power consumption, and cooling requirements.

The operating wavelength of the laser designator or illuminator depends upon several critical factors. A laser illuminator operating at 1.064 μm will experience higher extinction coefficient for related smoke agents ranging from 3.64 km^{-1} for fog to 2.19 km^{-1} for smog. However, a carbon dioxide laser operating at 10.6 μm will have the lowest extinction coefficient ranging from 0.047 km^{-1} for fog to 0.152 km^{-1} for acid fog. A CO_2 laser designator operating at 10.6 μm has the following advantages over the solid-state neodymium-doped yttrium aluminum garnet (Nd:YAG) laser designator operating at 1.064 μm:

- It offers the lowest extinction coefficient.
- It offers several operational advantages such as lower attenuation in fog and haze.
- It provides better penetration capability in fog and smog.
- It provides conversion efficiency better than 20%–25% compared to 1%–5% for the Nd:YAG laser.
- It offers optimum safety of eyes.
- It provides diffraction-limited beam width that offers least vulnerability to IR countermeasures.
- It offers greater receiver sensitivity.
- Despite all these benefits, the 10.6 μm laser illuminators suffer excessive weight, size, cost, and system complexity. CO_2 lasers have been deployed for a long time for airborne military applications. Detection range capability of a CO_2 radar under different weather conditions is shown in Figure 3.6. As mentioned earlier, this particular laser is widely used as an airborne radar and as a CW laser illuminator.

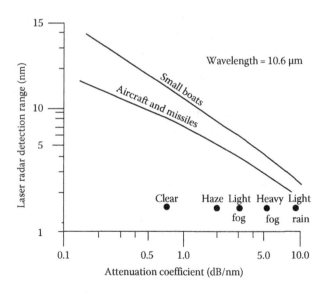

Figure 3.6 Detection range capability of a CO_2 laser radar under different weather conditions.

Laser Ranging System for Precision Weapon Delivery

Studies performed by the author on weapon delivery accuracy reveal that a laser-based ranging system offers the highest air-to-ground ranging accuracy, which is very important for precision weapon delivery capability and for achieving optimum kill probability. The studies further reveal that most ranging systems must have a ranging accuracy of a few meters ranging from 5 to 7 m out to 20 km, which is considered adequate for weapon delivery against most tactical targets. Note that under ideal conditions, even the naked eye has an angular error of about 0.5 mm rad or 3 ft at a range of 1 nmile. However, a laser operator reading errors could be degraded due to pulse rise variations, trigging circuit jitters, quantization error, and optical dome boresight error. The ranging error is a function of slant range, depression error, and angular error of the sensor. Laser sensor error for a sensor is less than 1 m rad at an IR operating wavelength of 1.064 μm. Calculated values of ranging errors from a laser ranging sensor aboard a UAV are shown in Table 3.6.

Electro-Optical Guided Missile

An electro-optical guided missile (EOGM) is also known as an IR guided missile. This missile can use either spot detector technology or image detector technology. This missile deploys an IR seeker that seeks the IR signatures originating from hot surfaces or exhaust gases coming out from the rear of jet engines. The spot detector uses a spot on the target or heat source emitting IR energy. A sophisticated EOGM uses exotic techniques such as optical beam rider, which is

Table 3.6 Calculated Laser Ranging Error as a Function of Slant Range and Depression Angle (ft)

	DEPRESSION ANGLE (DEG)				
SLANT RANGE (NM)	20	30	40	50	60
1	15	10.5	7.2	3.1	1.8
2	30	21	14.4	6.0	3.5
5	75	52	36	15	9.1
10	150	105	72	30	18

also known as a beam rider technique. The laser designator provides the guided signals. The seeker optics and the missile scanning mechanism provide directional information to the control electronics. Using the electronic gates, the operator can designate the target and fire the missile for a direct hit on the target.

IR Lasers to Counter the IR Missile Threat

The current or next generation of UCAVs might face IR missile threats. To eliminate such threat, high-energy IR sources or IR lasers are required. Multiple high-energy laser (HEL) sources are possible using long-wavelength IR laser technology such as HF lasers emitting at 2.8 μm, DF lasers emitting at 3.8 μm, CO_2 lasers emitting at 10.6 μm, or COIL lasers emitting at 1.315 μm. Before selecting a specific type of high-power IR source to contain a specific IR threat, it is essential to undertake trade-off studies on the weight, size, and power consumption requirements. Studies performed by the author seem to prefer CO_2 laser technology for the ECM technique, provided weight, size, and power consumption will justify the adoption of such a technique to counter tactical theater IR missiles.

Studies performed by the author seem to indicate that longitudinally excited CO_2 lasers using high-voltage pulses offer many orders of improvement in output power and energy levels. Transverse excited atmospheric (TEA) laser transmitters are capable of producing several MW of pulsed output levels with conversion efficiencies ranging from 15% to 25%. It appears that TEA IR lasers are relatively simple and compact and are best suited for UCAV applications, provided weight and size parameters will justify a viable IR laser transmitter source.

Diode-Pumped Solid-State IR Lasers

A direct high-power IR source with minimum weight and size is not feasible with current technology. However, the DPSS laser technology offers some hope. This particular technology allows several diode bars or arrays to achieve several hundreds of mW of CW power levels at room temperature. Research studies performed by various solid-state scientists indicate that DPSS lasers offer power levels with improved

Table 3.7 Performance Comparison between Diode-Pumped
Solid-State and FPSS Laser Designs

PERFORMANCE PARAMETERS	DPSS LASER	FPSS LASER
Electrical-to-light efficiency (%)	50	70
Overall efficiency (%)	10	12
Size	Small	20% larger
Weight	Light	25% heavier
Reliability or mean time between failures (MTBF) (h)	10,000 (min)	1200
Beam quality	Excellent	Moderate
Amplitude stability	High	Moderate

efficiencies compared to flash lamp solid-state laser technology. A technical article published in 1999 has demonstrated that a DPSS IR laser using Nd:YAG diode arrays has generated a peak power of 4 kW and a continuous wattage of 240 W with aperture dimensions of 1×2 em and with a beam dimension of 25×280 m rad. Solid-state scientists predict that a peak power close to 10 kW and CW power exceeding 500 W are now possible using the same technology. DPSS lasers using Nd:YAG diode arrays offer high brightness and high average power levels with conversion efficiencies greater than 10% with no cryogenic cooling. However, the efficiency could approach 15% with cryogenic cooling, but at higher cost, weight, size, and complexity. Performance comparison between DPSS and FPSS laser designs can be seen in Table 3.7. The emission wavelength of both lasers is approximately 1.064 μm. Solid-state lasers with an emission wavelength of 2 μm and with higher total differential quantum efficiencies (TDQEs) are available. Specific values of TDQEs as a function of reflectivity and laser chip size are shown in Figure 3.7.

Other Types of Lasers Available but Maybe Not Suitable for UAV Applications

Dye lasers and optical fiber lasers capable of operating over a wide tunable range are available but are not suited for current or future UAV applications. COILs suffer large weight and size and hence are not suitable for UCAV applications. TEA lasers are available and have been deployed in airborne military applications, but they could be suitable for current and next generations of UCAVs or UAVs, if their

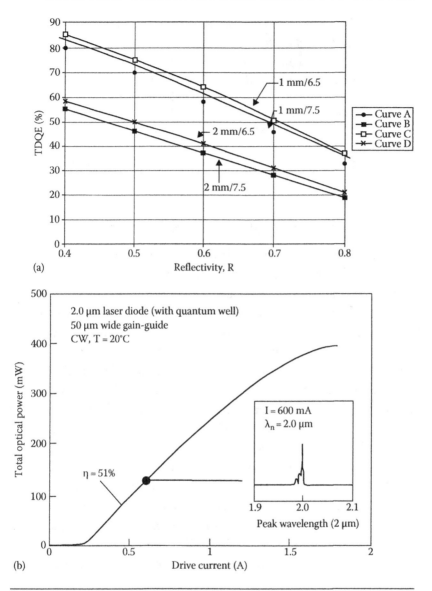

Figure 3.7 (a) Total differential quantum efficiency and (b) optical power for a 2 µm semiconductor laser.

weight and size reduced to acceptable limits. Note that the efficiency of high-power TEA lasers could be as high as 25%. This particular laser has been deployed in tracking long-range missiles with high probability and tracking accuracy. Performance characteristics of CO_2 TEA lasers are briefly summarized in Table 3.8.

Table 3.8 Characteristics of Transverse Excited Atmospheric Laser

PERFORMANCE CHARACTERISTICS	
Operating wavelength (μm)	10.6
Pulse power capability (kW)	500 (max)
Pulse width (ns)	50 or less
Pulse repetition frequency (PRF) (Hz)	1–5
Beam divergence (m rad)	<0.5
Range resolution (ft)	<25
CO_2 gas refill requirement	After every 2 million shots

Space Communication Laser System Employing Rare Earth Materials

The space communication laser involves three distinct communication channels, namely, space-to-space communication channel, space-to-aircraft communication channel, and space-to-ground communication channel. Note that each communication channel has its own specific performance requirements. Development of the Nd:YAG laser was initiated by the U.S. Air Force in 1971 with the development of specific components. This development has progressed from basic component development through breadboard system and breadboard system demonstrations to engineering feasibility model (EFM) system space usable concepts, simulated 40,000 km intersatellite operation, and certain available flight-qualified components.

Space flight tests were made on the transmitter package while in orbit with a transfer of date at the rate of 1000 Mbits/s to a ground station. Later on, a satellite was launched into an orbit with its apogee at synchronous altitude, so that the nominal task link range R will be close to 40,000 km or 19,320 nmiles. The purpose of flight tests was to demonstrate several critical system features as described in the following:

- Automatic acquisition and alignment of the two widely separated terminals by laser means
- Demonstration of intertracking accuracy terminal to a ±1 μrad (peak value)
- Transfer of data at the rate of 1000 Mbits/s
- Verification of system operation and reliability in space environments

- Evaluation of the downlink as a function of metrological parameters
- Communication system performance as a function of components variances
- Communication system performance as a function of laser beam variations

Important features of this space laser communication system include

- *Type of satellite deployed*: Synchronous satellite
- *Laser wavelength*: 0.532 μm
- *Communication range*: 40,000 km
- *Angular tracking error*: 1 μrad
- *Laser output (CW)*: 270 (mW)
- *Communication link range*: 19.320 nmiles
- *Date transmission rate to ground station*: 1000 Mbits/s
- *Electrical-to-optical efficiency*: 1% (max) at room temperature, which can be increased using cryogenic technology

In summary, the Nd:YAG communication system is a direct detection system, whereas the CO_2 communication system uses heterodyne detection technology, which requires a stable local oscillator in the receiver package. The stability and the reliability of the oscillator are of paramount importance. Furthermore, the space-to-earth communication channel requires a high-data-rate laser transmitter in synchronous orbit transmitting to an earth-based communication receiver.

Forward-Looking Infrared Sensors

In this section, two IR sensors will be described as playing a critical role in detection, recognition, and tracking of targets. One sensor is called the FLIR sensor, while the other sensor is known as the IR search and tracking (IRST) sensor. The FLIR deploys cooled and uncooled detectors, while the sensor uses cryogenic technology to cool the mercury-cadmium-tellurium (HgCdTe) detectors that can be cooled down to a cryogenic temperature of 77 K to achieve a high-quality image of the target. These sensors will be briefly described.

Forward-Looking Infrared Sensors for UAV Applications

FLIR sensors are widely used in detection, tracking, and recognition of targets in the forward sector views. Different FOVs are assigned for target detection, track, and recognition. This sensor offers three distinct FOVs, namely, narrow FOV, moderate FOV, and broad FOV. The broad FOV is recommended for target search and detection, the moderate FOV is designed for target tracking, and the narrow FOV is intended for target identification and for target classification. These sensors usually deploy the integration of cryogenically cooled FPAs. Battery-powered, Stirling microcoolers are used to cool the HgCdTe-IR detectors to reduce significantly the dark current and the detectivity. Note that the detectivity is strictly dependent on cryogenic cooling and the processing gain. The Stirling microcoolers weigh less than 11 oz, consume electrical power less than 3.5 W, have an operating life exceeding 2000 h, and take about 4 min to cool from room temperature to 77 K. Such sensors are most ideal for UCAV and UAV applications, if target detection, acquisition, and identification are the principal requirements of the UCAV or UAV platform when operating in military conflict regions. Note that the FLIR sensors are light, compact, and inexpensive and are widely used for surveillance and reconnaissance missions. The two real components of an uncooled FLIR are the detector and the optics. Note that the uncooled FLIRs are less sensitive, very light, and require the least electrical power. Uncooled FLIRs are best suited for short-range missions. However, the cryogenic cooling results in greater sensitivity and excellent performance over long operating ranges. The latest cooled FLIR designs offer long battery life and better long-range performance.

In summary, it can be stated that the FLIR sensor is the most critical sensor for the UAV platform because it provides a complete video picture of the lower forward sector ahead of the UAV needed to locate and identify the ground targets in poor visibility environments. Furthermore, the FLIR is of critical importance in accomplishing specific military missions under all weather conditions. As mentioned earlier, this sensor provides precision navigation and weapon delivery capability virtually in all weather environments. It offers to IFOVs a wide FOV for target search and navigation, and narrow FOV for target track and weapon delivery. This particular sensor is comprised of four line replacement units (LRUs) as described in the following:

- FLIR optical assembly mounted on a gyro-stabilized platform
- Electronics module containing all electronic circuits and devices and cryogenically cooled detector array
- Power supply unit with minimum voltage ripples
- Processing assembly

IRST Sensor for UAV Deployment

This particular IRST sensor is the most important for undertaking search and tracking missions in the front sector. The sensor contains the most sensitive receiver, which performs the integration of cryogenically cooled micro-FPAs for optimum sensitivity and consistent sensor performance in military environments. This sensor has demonstrated a sensitivity better than 10^{11} em $(Hz)^{0.5}$ per watt at a cryogenic temperature of 77 K. This particular "infrared search and track" sensor is equipped with time delay integrating (TDI) circuits, which will significantly improve the tracking accuracies better than 1 m rad.

Performance Capabilities and Limitations of IRST Sensors This sensor is capable of detecting and tracking decoys, aircraft, and cruise missiles at long distances with high resolution and high altitudes. When installed on a UCAV at an appropriate location, the sensor can provide detection ranges close to 100 mi under clear weather conditions. The overall performance of the sensor is dependent on several parameters, such as dwell time, number of detectors, cryogenic cooling temperature, detector sensitivity, photon quantization, and IFOV of the detector array. Since this is a passive airborne sensor, longer detection and racking ranges are possible with minimum power consumption, weight, and size compared to an active sensor.

Performance parameters of the IRST sensor can be briefly highlighted as follows:

- Offers detection and tracking ranges of airborne targets in excess of 100 mi.
- Provides raid-count information in the presence of jamming.

- Yields high detectivity when the HgCdTe detectors are cooled down to a cryogenic temperature of 77 K.
- Provides early warning against a hostile fighter aircraft equipped with air-to-air missiles and cruise missiles.
- The sensor performance is typically optimized over 3–5 and 8–12 μm spectral ranges, depending on the jet engine thrust generating level, exhaust temperature, UAV engine, and operating altitude. It is extremely important to point out that a 2-D, cryogenically cooled detector array provides optimum sensor performance for both the staring and scanning detector assemblies.
- The FPA architecture consists of HgCdTe detector arrays coupled directly to an array of silicon charge-transfer devices (CTDs). When the IR radiation is incident on the detector array, it is converted into an electrical signal, which is immediately coupled into a storage device directly beneath the detector. A set of pulses are applied to the readout register, which transfers this charge from one row of the array to a preamplifier. These steps are repeated until all rows of the shift register have been read to the signal processor, providing a single-frame image of the IR scene. Incorporating both the detector and the CTD arrays on to one substrate material will yield optimum sensor performance with minimum weight, size, and cost [6].
- A system architecture incorporating several staring FPAs, each FPA having a wide FOV, offers the UAV or UCAV platform detection capability over full 360° spherical coverage against the antiaircraft missile improvement in the sensor performance.
- *Typical wide FOV:* 30° × 20°.
- *Typical medium FOV:* 20° × 15°.
- *Typical narrow FOV:* 5° × 5°.

Laser Range Finder Sensor for UAVs: The primary function of this sensor is to provide the most accurate range-finding information with maximum safety to the operator. A multifunction laser range finder (LRF) is also available, which offers various functions needed for successful execution of military missions with minimum weight, size, and power consumption, which are the principal design requirements of the sensors. The following laser wavelength requirements can be recommended for various sensors for UCAV platforms:

- 1064 nm for laser decimeter and range finder
- 1500 nm for eye-safe laser designator
- 1525 nm for eye-safe LRF
- 3000–4000 nm or 3–4 μm for IR jamming applications

Studies performed by the author on laser-based range finders indicate that an operating range of 2–3 km is quite appropriate for an Nd:YAG eye-safe LRF for a battlefield tank or space application, where accuracy, power consumption, weight, size, and power consumptions are of paramount importance. Typical performance parameters of LRF sensors can be summarized as follows [7]:

- *Laser transmitter power*: 10 W
- *Laser operating modes*: CW or Q-switched pulsed
- *Number of detectors*: 12
- *Receive optic size*: 5 in
- *Instantaneous FOV*: 0.2–0.3 m rad
- *Imaging FOV*: 12 (AZ) × 10 m rad (EL)
- *Maximum operating range*: 2.75 km
- *Estimated sensor weight*: 40 lb (maximum)
- *Power consumption*: 350 W (maximum)

IR Detectors for UAV Applications: There are various targets that yield pronounced IR signatures at specific IR wavelengths. Essentially, it means that different types of IR detectors are needed to respond most effectively to specific wavelengths. Prediction of the IR radiation level and emitting wavelengths from specific objects or IR sources will be briefly discussed. Studies performed by the author on various IR detectors reveal that some IR detectors are very responsive over a specific IR spectrum and such detectors that are most appropriate will be discussed with emphasis on spectral range, responsivity, and temporal response.

Classification of IR Detectors: IR detectors are available in three distinct broad classifications, namely, noncryogenic detectors, cryogenic detectors, and FPA detectors. Cryogenic detectors require cryogenic cooling to meet specific detector performance requirements such as detectivity and responsivity. Note that optical detectors are classified in different categories such as high-speed detectors, low-power detectors, high-power detectors, pulse energy detectors, time-domain detectors, frequency-domain detectors, and photon detectors.

Estimation of IR Signature from Airborne Targets: IR signatures from airborne moving targets can be estimated but not accurately [6]. Aircraft or missile skin temperature is a function of aerodynamic shape, speed, operating altitude, and atmospheric conditions. As the speed increases, the surface or skin temperature increases due to laminar flow over the body surface. Skin temperature of the missile or aircraft flying above 37,000 ft can be estimated by the following equation:

$$T_{skin} (K) = [216.7(1 + 0.16 \ M)] \tag{3.1}$$

where

T_{skin} is the skin temperature in Kelvin
M is the speed of the missile or aircraft

Skin temperature and the wavelength of the peak IR radiation as a function of speed are given in Table 3.9.

Table 3.9   IR Radiant Emittance over a Specified Spectral Bandwidth

TEMPERATURE (K)	SPECTRAL BANDWIDTH (μM)	% OF RADIANT EMITTANCE
500	6.00–5.00	11.3
1000	3.00–2.00	20.4
1000	3.00–2.25	16.3
1000	3.00–2.50	11.2
2000	1.500–1.00	20.6
2000	1.50–1.20	13.3
2000	1.50–1.25	11.2
3000	1.00–0.50	26.1
2000	1.00–0.75	16.4
2000	1.00–0.85	10.1

Source: Jha, A.R., *Infrared Technology: Application to Electro-Optics, Photonic Devices and Sensors,* John Wiley & Sons, Inc., New York, 2000, pp. 105–130.

From the data presented in Table 3.9, it appears that at a specific skin temperature at 1000 K or more, there could be at least three spectral bandwidths with three distinct radiant levels. Furthermore, the skin temperature and the IR radiant level will change as a function of speed. One will see that the constants mentioned in the mathematical expression will change with the speed and operating altitude. Note that the tabulated values are based on certain assumptions, and therefore, the author will consider these values as estimated values with errors ranging from 5% to 10% (minimum).

Impact of Aircraft Maneuver on the IR Signature The IR signature of an aircraft or a missile under static conditions is quite different from the IR signature under dynamic conditions. When the aircraft is flying at a certain speed and is under yaw, pitch, or roll conditions, its IR signature will be radically different. Computer simulations performed based on certain assumptions indicate that the maximum IR radiation level from a jet aircraft occurs at its tail aspect, whereas the minimum IR radiation intensity occurs at the nose aspect as shown in Figure 3.8 with and without atmospheric effects. It is evident from this figure that the peak intensity of 1000 W/sr at the tail aspect is reduced to 450 W/sr at the nose aspect. Note in the forward flight sector that the IR radiation levels are close to negligible. In addition, impact of roll and pitch angles on the radiant intensity can be seen from the curves as shown in Figure 3.8. It is interesting to state that the radiant intensity experiences significant reduction due to atmospheric attenuation irrespective of the aircraft maneuvering angles.

One can see from Figure 3.8 that the reduction in the IR intensity level is relatively moderate for pitch angles ranging from +10° to −20° at aircraft altitudes exceeding 5000 ft and at slant ranges less than 5000 m. Computer simulation data obtained by the author reveal that maximum intensity occurs over the 2.8–3.2 µm spectral region with tailpipe temperatures ranging from 900 to 1050 K. However, when the jet aircraft is operating in the afterburner mode, the maximum radiation intensity will be roughly 30%–40% higher due to higher tailpipe temperature and thrust level over the 1.3–1.8 µm spectral region. When a jet aircraft is operating in an afterburner mode, the aircraft must complete its assigned mission in minimum time to avoid lock-on by enemy IR missiles.

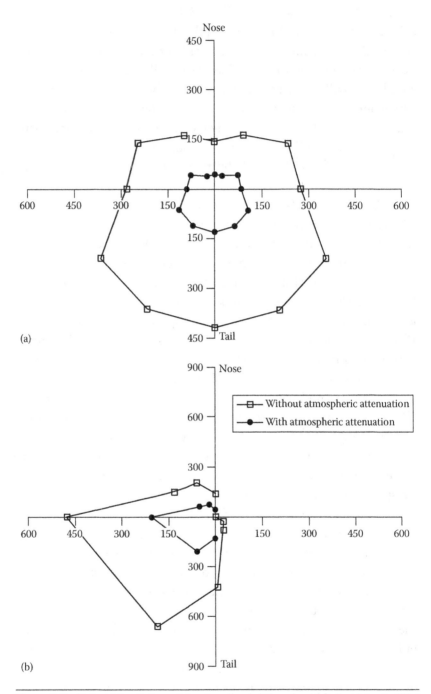

Figure 3.8 Radiant intensity from a jet engine roll and pitch angles. (a) Radiant intensity (W/sr) under roll conditions and (b) radiant intensity (W/sr) under pitch conditions.

Types of Infrared Detectors

Various types of optical detectors are available to measure parameters either in the frequency domain or in the time domain. Once the domain is known, the detector material is selected to provide optimum sensitivity. The selection of a detector is strictly a function of type of application, operating mode (CW or pulsed), and measurement of parameters involved. Semiconductor materials such as Si, PbS, PbS, PbSe, InGaAs, and HgCdTe are widely used in the design of IR detectors because they offer acceptable performance at room temperature with minimum cost and complexity. Only a few detectors will be discussed in detail even though several types of detectors are available to measure the IR response, which can be summarized as follows:

- Time-domain detectors
- Frequency-domain detectors
- Low-power, high-speed detectors
- High-speed detectors
- Semiconductor photovoltaic cell detectors
- Photon counting detectors
- Quantum detectors
- FPA detectors (uncooled or cooled)

Description and Performance Capabilities of Most Popular IR Detectors   In this section, only the most widely used IR detectors will be discussed with emphasis on detectivity, responsivity, and sensitivity parameters of the IR detectors. Performance parameters of selected types of IR detectors will be described in great detail. Attempt will be made to focus on one in each category of the IR detector. The Johnson noise or the thermal noise in the IR detector is strictly due to the temperature rise in the junction of the detector element, which can be reduced at cryogenic temperatures. However, use of cryogenic cooling will increase the cost and complexity. The author will now focus on the IR detector that will not require cryogenic cooling except certain detectors.

Photon Detectors   Detector theory indicates that photon detectors have more photon-generated carriers than the thermally generated carriers.

Table 3.10 Photon Detectors Operating at Various Temperatures

PHOTON DETECTOR MATERIAL	OPERATING TEMPERATURES (K)
PbS	300, 193, 77
Si	300
InAs	300, 195, 77
InSb	300, 77

Note that a solar cell is a photon detector. The most widely used photon detectors are summarized in Table 3.10 with or without cryogenic cooling.

Low-Power, High-Speed IR Detectors   Note that in ordinary detectors, coherent light causes reading errors across the detector surface, which causes error rates ranging from 5% to 8%. These errors should be significantly reduced to achieve uniform and stable detector response. A built-in micro attenuator provides low refection, high damage threshold, spectral flatness, and accurate low-power IR measurements. Note that the spectral responsivity of such detectors can be optimized for a specific spectral band for impulse response. These detectors can be optimized for narrow and wide UV bands and for broadband IR response.

High-Power, High-Speed IR Detectors: High-speed, high-power detectors are designed to provide fast response with low spectral response, high damage threshold, flat response over the 1–12 μm spectral range, and CW or pulsed signal capability. Such a detector involves isothermal disk design, which offers ultrahigh accuracy, uniformity, and reliability during the measurements of high-power CW or pulsed signals. Miniaturized pyroelectric, high-power detectors come in T0-18 packages with thermally stable coating, which offers unfiltered IR response over the 1–15 microsecond period. High-power detectors are widely used for power measurements for high-power Nd:YAG, Ti:sapphire, and holmium IR lasers.

The high-speed detector design involves large-area fiber coupling, high conversion gain, and broad spectral response, which are possible for InGaAs using the metal–semiconductor–metal design technology. This particular detector is widely used in the detection of high-speed digital and microwave signals with sensitivity ranging from 0.2 to 0.8 A/W over the 400–1700 nm spectral range and efficient high-speed optical-to-electrical conversion with temporal response from 15 ps at 30 GHz to 30 ps at 15 GHz.

Typically, the high-speed detector configuration includes lithium batteries, each with internal voltage regulation capability, and a battery monitoring device that lights up a light-emitting diode (LED) indicating that battery replacement is needed. This detector offers a 12 ps full-width, half-maximum (FWHM) impulse response, fast response with minimum ringing, bandwidth up to 60 GHz in the visible spectral region and 45 GHz in the near-IR region, and high responsivity over the 1310–1550 nm spectral range.

High-Performance IR Detectors for UCAV Sensor Applications: Stable performance, high reliability, optimum sensitivity, and fast and accurate response are the principal requirements for the IR detectors used by the IR line scanner (IRLS), FLIR sensor, and the IRST system aboard the UCAV. Note that these IR detector performance parameters are possible using certain junction materials and at optimum cryogenic temperatures. Studies undertaken by the author [6] on IR detectors reveal that germanium and germanium–mercury detectors require operating temperatures well below 30 K for optimum detector sensitivity, while the HgCdTe and PbSn detectors yield optimum performance at 77 K. The ternary alloy detectors such as HgCdTe and PbSnTe are widely used over the 8–14 μm spectral range because of their improved responsivity and directivity at the cryogenic temperature of 77 K. That is why HgCdTe detectors are widely used by the IRLS, FLIR, and IRST sensors.

Quantum IR Detectors: Quantum IR detectors are generally used in the visible and near-IR spectral regions. For optimum detector performance, cryogenic cooling is essential. Cryogenic cooled quantum detectors are best suited for fiber optic–based systems, such as optical data links and fiber optic ring lasers. Performance parameters of these detectors such as sensitivity, detectivity, responsivity, response time, dark current level, and noise-equivalent power (NEP) will significantly improve at cryogenic temperatures. However, quantum IR detectors have to be cooled below 77 K to achieve the background-limited performance level.

Photomultiplier Tube Detectors: Photomultiplier tube (PMT) is the most efficient photon counting device because of its high inherent gain and low noise factor. In nuclear imaging, it is the PMT that creates high image quality and outstanding detector performance because of high gain. These devices are deployed by the armed forces in conjunction with night vision sensors in combat regions, particularly when the ambient light conditions are very poor. Critical elements of a PMT system can be listed as follows:

- Photocathode.
- Electro-optical collection lens.
- First dynode.
- Second dynode.
- Third dynode.
- Fourth or last dynode.
- Collector collects the increasing number of secondary electrons generated by four dynodes or any number of dynode stages.
- PMT can generate one million or more electrons.
- *The transit time through the PMT* is 15 ns.
- *The pulse width can be as short as* 1.5 ns.
- *The peak output pulse voltage* is measured in mV.
- *The pulse current* is 100 μA.
- Pulse discriminator.
- Histogram known as pulse height distribution (PHD).
- Histogram development requires a collection of millions of pulses and storing them in a memory.
- Its entering light level is as low as few femtowatts (10^{12} W).
- Efficiency and reliability of a PMT used in military field environments must be given prime consideration.
- A 2% change in the power supply voltage will change the PMT gain by 15%.
- DC power supply fluctuations must be kept below 1% to maintain high accuracy.
- Each element of the lens array is made up of two spherical lenses.
- It has the telescope ability to collect the light with incident angles less than 100 m rad or 5.68°.
- *The PMT detector efficiency* is 60% (minimum).

Optical Detectors   NEP is the most critical performance parameter of an optical detector. A high-temperature, transition-edge bolometer made from a superconducting yttrium barium copper oxide (YBCO) film with a critical temperature of 91 K on the strontium titanate substrate provides an NEP as low as 7×10^{-12} W/(Hz)$^{0.5}$, which is impossible to achieve from a room temperature IR detector for radiation measurements at emitting wavelengths exceeding 13 μm.

IR and Television Cameras

Scientific and strategic studies performed by the author on the UAV sensor requirements seem to reveal that high-resolution thermal IR technology and a miniaturized, high-definition television sensor play critical roles in precision navigation and target tracking in combat environments. The miniaturized thermal IR cameras are designed specifically to significantly improve the surveillance and reconnaissance capabilities in combat environments. These thermal IR cameras are built to operate under extreme operational conditions. The IR cameras deliver state-of-the-art performance over-the-hill reconnaissance and surveillance functions, close-range attack capabilities, and remote location identification, classification, and tracking of hostile targets. Note that the high-definition TV sensor offers a clear picture of the critical scene in the conflict zone with minimum cost, which is not possible otherwise.

Performance Capabilities of Various Gyros for UAV Navigation

A gyroscope (gyro) plays a critical role in the navigation of aircraft, UAV, helicopter or space traveling systems. A gyroscope contains multiple small components and therefore is known as a multicomponent structure. Each of these components has different statistical characteristics. When measuring angle orientation, the gyro signal should be integrated with its allied components. These noise components have different rates of error accumulation during the integration process because the error accumulation is dependent on the statistical characteristics of noise. It is important to note that white noise does not affect the attitude error significantly, but random noise does.

In addition to noise error, there is a problem of determining gyro drift components' statistical parameters to characterize gyro accuracy. The essence of famous Allan variance is comparatively a very simple method and allows estimating noise-level intensities of different nature. The essence of this method is to increase the averaging time of the gyro drift measurement and generate a graph of the root square of averaged drift variance called standard variation versus averaging time. In the Allan standard graph, the presence or absence of various error components can be seen. The errors are shown on the Allan graph. Note that most of the errors shown in such a graph are inherent to almost all modern gyros. This graph allows detecting different error components that exist in the measurement data. Note that the total Allan variance can be written as the sum of variances of each gyro component. For specific details on the Allan graph, one should refer to a standard textbook.

Most Popular Gyros Deployed by Aviation Industry

Comprehensive studies undertaken by the author on the subject concerned indicate that the most popular gyroscopes used by various aircraft, UAVs, and helicopters can be briefly summarized as follows:

- Ring laser gyros (RLGs)
- Fiber optic gyros (FOGs)
- Coriolis vibratory gyros (CVG)
- Microelectromechanical system (MEMS)-based gyros

Performance Summary for Various Types of Gyros Several types of FOGs have been developed for applications in commercial, military, and space vehicles. Studies performed by the author on RLG devices reveal that a passive device that uses a finesse fiber ring resonator concept offers better performance with a much shorter optical fiber loop than an interferometer-type FOG. This particular type of gyro is best suited for fighter aircraft, missiles, space sensors, and UAVs, where weight and size are the demanding requirements. Note that bias is induced due to the optical Kerr effect contributing to a dominant noise source. This bias produced is proportional to the difference in intensity between the CW and coded-CW light waves present in the resonator. Note that even a slight imbalance between the two light waves can produce a bias error greater than shot noise. Note that the Kerr effect–induced drift can slightly degrade the gyro performance. The studies further reveal that the Kerr effect–induced drift is less than 0.1 μrad/s.

Studies performed on gyros seem to indicate that most of the gyros suffer errors and that is why different types of error components can be expected in measurement of performance data. The research studies further indicate that the metallic resonator manufacturing cost is much less than that of the device made of quartz or optical resonator RLG and fiber optic coil with polarization-maintaining fiber for moderate- and high-accuracy FOG. On the other hand, a CVG assembly sensing element is simple and takes less money and a short manufacturing time. As far as the reliability of CVG is concerned, it is 10 times more than that of RLG and FOG devices.

MEMS-based gyros [8] are best suited for airborne system applications, where minimum weight and size, precision navigation, and ultrahigh reliability are the principal design requirements. Note that the highest precision inertial measurement units (IMUs) such as RLGs and FOGs have been in use for several years and they provide satisfactory performance but at the expense of high cost, excessive power consumption, and large packaging assembly. In summary, the MEMS-based gyros must be used in applications where precision guidance and most accurate navigation are the principal performance requirements.

Summary

This chapter deals with EO and RF connectors, devices, and sensors widely used by UAVs and UCAVs. Because of deployment of these vehicles in hostile environments, maximum emphasis is placed on reliability, maintainability, longevity, and structural requirements. Performance requirements including electrical and mechanical requirements are specified for the EO and RF components with particular emphasis on weight, size, RF leakage, and mechanical integrity. Critical requirements for EO and RF cables are specified, namely, IL, input VSWR, and protection layer under severe operating conditions. Performance requirements for the passive RF components and devices are briefly summarized with emphasis on weight, size, and power consumption.

Performance characteristics of the micro- or NANO-SAR for UAV applications are briefly summarized with emphasis on weight, physical dimensions, and power consumption. The estimated values of these parameters are 1.0 lb, 5.5 × 3.5 × 2.2 in, and 25 W, respectively. This NANO-SAR offers a resolution better than 1 to 4 ft. Deep-drawn microparts with surface finish better than 120 µin. are used in the fabrication of SAR and sensitive receivers with no RF energy leakage provision. Stringent specifications are provided for RF connectors and cables that are needed for the tactical data link to ensure the highest security and integrity in the transmission of secured data.

Performance requirements for semiactive microwave devices such as YIG filters, solid-state varactorbased limiters, and ferrite-based limiters are defined with particular emphasis on weight, size, power consumption, passband IL, limiter threshold, dynamic range, and stop-band rejection. Stringent performance requirements are specified for the reconnaissance and surveillance receivers with particular emphasis on noise figure, instantaneous bandwidth, dynamic range, weight, size, and power consumption.

Advantages of RF amplifiers using MMIC technology are highlighted for the benefits of UAV integration engineers and readers. Other RF amplifiers are identified including uncooled parametric amplifiers and MMIC amplifiers, which yield state-of-the-art performance with minimum weight, size, and power consumption. The MMIC amplifier offers AM-to-PM conversion performance better than ±2.2° and amplitude tracking better than ±0.5 dB. Note that MMIC amplifiers are available in various RF bands ranging from S-band to Ka-band. These amplifiers are best suited for UAV applications where unstable flight conditions, harsh thermal environments, and severe mechanical conditions can be expected any time. MMIC amplifiers offer optimum design flexibility and can deploy any transistor type ranging from MESFET to HEMT to p-HEMT devices.

The author has identified failure mechanisms in discrete elements used in the design of RF oscillators, RF amplifiers, YIG-tunable filters, YIG oscillators, solid-state limiters, ferrite limiters, and C lasers and TEA lasers. Studies undertaken by the author on lasers seem to indicate that the CO_2 laser is best suited for laser illuminators or target designators for laser-guided missiles due to its superior beam stability, beam circularity, and highest laser efficiency ranging from 20% to 30%. Electrical performance of the TEA laser is summarized with emphasis on efficiency, weight, size, and power consumption. Note that the TEA laser offers an excellent beam divergence capability of 0.5 μm, range resolution better than 25 ft, and pulse width less than 50 ns. Laser seekers operate over a 3.2–4.3 μm spectral range to minimize false alarms from the clouds. Note that the low-power Nd:YAG laser operating at 1.064 μm is best suited for short-range laser operation. Furthermore, that the Nd:YAG-based LRF is widely used for ranging applications, where range accuracy is of a prime consideration. It is necessary to mention that use of cryogenically cooled HgCdTe detector arrays in FLIR and IRST sensors is absolutely essential if optimum performance of these sensors is desired. Performance parameters of various IR sensors including FLIR sensors with three appropriate FOVs and cryogenically cooled IRST sensors for possible deployment in the next generation of UCAVs are summarized. Performance capabilities of MEMS-based gyros as shown in Figure 3.9 are discussed

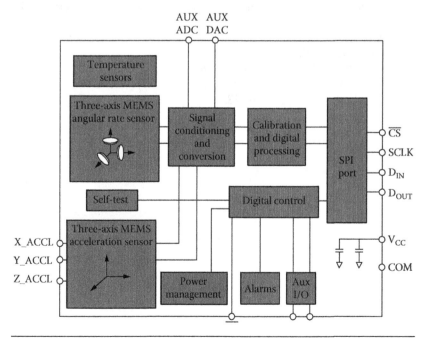

Figure 3.9 Block diagram of a MEMS-based gyro system showing the critical and functional elements.

in great detail with particular emphasis on reliability, navigation accuracy, stability (few tenths of a degree per hour), power consumption, weight, and size. This particular gyro is most ideal for application, where an inertial sensor with six degrees of freedom is required. Note that the MEMS-based gyro is best suited for UAV applications.

References

1. IMSAR, Electrical, RF sensors, and components for UAV, IMSAR LLC, Springville, UT, 2014.
2. A.R. Jha, Technical progress report: X-band solid state limiter using high-Q varactor diodes, Jha Technical Consulting Services, Cerritos, CA, July 1965, pp. 8–18.
3. M. Skolnik, *Radar Handbook*, McGraw-Hill Corporation, New York, 1970, pp. 12–29.
4. A.R. Jha, Technical report on MMIC amplifiers using HEMT and p-HEMT devices and operating at mm wave frequencies, Jha Technical Consulting Services, Cerritos, CA, March 2000, pp. 4–24.
5. A.R. Jha, *Monolithic Microwave Integrated Circuit (MMIC): Technology and Design*, Artech House, Inc., Norwood, MA, 1989, pp. 10–25.
6. A.R. Jha, *Infrared Technology: Application to Electro-Optics, Photonic Devices and Sensors*, John Wiley & Sons, Inc., New York, 2000, pp. 105–130.
7. A.R. Jha, *Cryogenic Technology and Applications*, Elsevier Academic Press, New York, 2006, pp. 217–221.
8. A.R. Jha, *MEMS and Nanotechnology-Based Sensors and Devices for Communications, Medical and Aerospace Applications*, CRC Press, New York, 2008, pp. 376–378.

4

UAV NAVIGATION SYSTEM AND FLIGHT CONTROL SYSTEM CRITICAL REQUIREMENTS

Introduction

This chapter focuses on navigation system and flight control system requirements for unmanned aerial vehicles (UAVs) and unmanned combat air vehicles (UCAVs). Comprehensive technical concepts are presented on development and practical implementation with major emphasis on system reliability and accuracy of navigation and automatic flight control systems (AFCSs). The navigation system and flight control system should be treated like twin brothers that cannot function or survive without the help of the other. Because of the tremendous increase of UAV activities in the sky, it is very important to develop a reliable and accurate navigation system and AFCS for security and survivability on the ground and in the air. The comprehensive review of research reports and technical papers by the author on the subject concerned seems to reveal that research scientists at the National Aviation University in Ukraine have developed a prototype navigation and practical implementation of required critical system elements. An integrated navigation system (INS) was designed, developed, and tested in a UAV with specific weight, size, and other relevant parameters. Published technical papers [1] indicate that the efficiency and functioning of the INS have significantly improved because of innovation techniques.

It is the view of aerospace scientists' innovation in UAV navigation, control, error-free algorithms, advanced processing technology, availability of appropriate materials, and compact propulsion system that will contribute to the development of the most efficient and safe UAVs. In the last decade, UAVs have been successfully exploited by most developed countries and a few underdeveloped countries in various applications, including firefighting, movie industries, monitoring of wild animals, communication and broadcasting services, digital mapping, and a host of other services including the country's defense and security.

For the safety of the vehicle and its contents and of the public in general, the accuracy and reliability of the navigation system and automatic flight system must be given principal consideration. The entire system consists of two parts, namely, orientation and navigation system and AFCS as illustrated in Figure 4.1. Note the orientation and navigation system deploys necessary sensors, electronics, and computing facilities in order to calculate the current UAV velocities, positions, and orientation parameters. The data generated are transferred to the AFCS. The flight control system solves the UAV flight trajectory information and provides inputs to the UAV surface actuators or mechanisms for the control of UAV platform flight.

Comprehensive research and development activities undertaken by various universities indicate that the development of UAV orientation and navigation systems should include the Global Navigation Satellite System (GNSS), the Inertial Navigation System (INS), and other essential sensors. In other words, an INS as described earlier will provide significant safety and reliability in UAV flight operations. It is interesting to point out that scientists at Ukraine's National

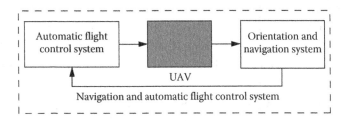

Figure 4.1 Block diagram showing the critical components of the unmanned aerial vehicle navigation and automatic flight control system.

Aviation University have done significant research and development work on autopilot systems and navigation systems for UAVs.

UAV Navigation System

Aviation scientists believe that the existing orientation and navigation systems should be upgraded using the latest technological concepts, if possible. Existing sensor designs should be redesigned or improved to achieve optimum safety and reliability of current and future UAVs. Note the upgraded or redesigned sensors can be used for other kinds of aircraft or transport. It should be noted that the UAV navigation system development and practical implementation must address the following critical items:

- Various types of sensors essential for UAV navigation
- Algorithm implementation with zero error
- Navigation system hardware
- Software development

After modifying or upgrading UAV sensors, algorithms, software, system hardware, and other critical subsystems, experimental tests must be conducted to validate the UAV navigation system. In the case of UCAVs operated by armed forces, additional sensors may be required, which are identified as follows:

- Microelectromechanical systems (MEMS)-based inertial measurement unit (IMU) with three-axis accelerometers
- Gyroscopes (gyros)
- Magnetometers
- Barometric altimeters
- GNSS receiver

Algorithms

Algorithms play a critical role in the design and operation of a commercial or military system. An algorithm is essentially a procedure for solving a complex mathematical problem. It can be treated as a step-by-step procedure for solving a problem or for finding the greatest common divisor. With reference to Chapter 4, two different types of algorithms may be required. For navigation calculations, it can be loaded into two categories: Strapdown Inertial Navigation System (SINS) functioning and correction calculations by GNSS and other appropriate sensors. It is necessary to state that SINS functioning requires four critical steps:

1. Correction of raw data for implementation of corrections
2. Altitude data updating as needed
3. Specific force transformations needed for navigation coordinate system
4. Velocity and position calculations needed for correct navigation path

Algorithms Appropriate for SINS Functioning

SINS functioning algorithms play an important role in dealing with various error corrections. Note that raw data require corrections in various bias levels, scale factors, and nonorthogonalities. The initial values of these parameters can be derived from laboratory experiments or field calibration procedures. Sensor measurements must be taken at some time interval Δt. In order to achieve precise digitization of required accelerations and angular rates, there should be an application of precise analog integration of high-grade IMU as a part of digitization. This means the outputs from each such system are required in incremental angles and incremental velocities due to specific force. Low-cost IMUs could provide angular rates and specific forces. It is necessary to use the following quadratic spline approximation for the calculations of incremental angles:

$$[\boldsymbol{\theta}_i] = (\Delta t/12)[5\omega(t_i) + 8\omega(t_{i-1}) - \omega(t_{i-2})] \qquad (4.1)$$

where
the vector quantity $\boldsymbol{\theta}_i$ represents the incremental angle
Δt is the interval
ω is the angular velocity
parameter i is greater than unity

The altitude update can be performed by the multiplication of the elementary quaternion, where quaternion is a complex number that is composed of one real number and three imaginary numbers.

This generates an amplitude quantity q representing the altitude update, which can be expressed as

$$q(t_i) = [q(t_{i-1})] \, [\delta q(t_i)] \qquad (4.2)$$

$$\delta q = [\delta \lambda_0(t_i) \; \delta \lambda_1(t_i) \; \delta \lambda_2(t_i) \; \delta \lambda_1(t_i)]^T \qquad (4.3)$$

This procedure can lead to a matrix form for the altitude update represented by the vector quantity $q(t_i)$. This matrix for $q(t_i)$ can be written as

$$q(t_i) = \left[\delta q_0(t_i) - \delta q_1(t_i) - \delta q_2(t_i) - \delta q_3(t_i) \right] [q_0(t_{i-1})]$$

$$\times \left[\delta q_1(t_i) \delta q_0(t_i) \delta q_3(t_i) - \delta q_3(t_i) \right] [q_1(t_{i-i})]$$

$$\times \left[\delta q_2(t_i) - \delta q_3(t_i) \delta q_0(t_i) \delta q_1(t_i) \right] [q_2(t_{i-1})]$$

$$\times \left[\delta q_3(t_i) \delta q_2(t_i) - dq_2(t_i) \delta q_0(t_i) \right] [q_3(t_{i-1})] \qquad (4.4)$$

Updated Euler angle expressions can be written as

$$\varphi = \arctan\left(2(q_2 q_3 + q_0 q_1) \big/ \left(q_0^2 - q_1^2 - q_2^2 + q_3^2 \right) \right) \qquad (4.5)$$

$$\theta = \arcsin(-2(q_1 q_3 - q_0 q_2)) \qquad (4.6)$$

$$\psi = \arctan\left(2(q_1 q_2 + q_3 q_0) \big/ \left(q_0^2 - q_1^2 - q_2^2 + q_3^2 \right) \right) \qquad (4.7)$$

After this, the specific forces are transformed to the navigation coordinate system using the updated quaternion and are corrected to gravitational and centripetal accelerations. Finally, velocity and position parameters are calculated using an approximation similar to the one used for incremental angle calculations and defined by Equation 4.1. Such calculations of velocity and position parameters decrease computational load while still providing the necessary computational accuracy.

Strapdown Inertial Navigation System (SINS) Algorithms

SINS correction signals are provided by the GNSS, magnetometer, and barometric altimeter using the Kalman filter equations. Note that the vector of SINS error is defined by a vector μ [1]. This vector represents the orientation error in velocity (v) and position (r) and is represented by vectors δv and δr, respectively. Now, SINS propagation error can be defined by the following expression:

$$\left[\frac{dx}{dt}\right] = [Fx + n] \tag{4.8}$$

where

$$x = \begin{matrix} \mu \\ \delta v \\ \delta r \end{matrix} \tag{4.9}$$

$$F = C \begin{matrix} 0 & 0 & 0 \\ 0 & 0 \\ 0 & I & 0 \end{matrix} \tag{4.10}$$

$$C = \begin{matrix} 0 & -a_3 & a_2 \\ a_3 & 0 & -a_1 \\ -a_2 & a_1 & 0 \end{matrix} \tag{4.11}$$

where

n represents the noise vector

0, I represent zero and eye matrices with dimensions equal to 3 × 3

vector a is a transport vector defined by $[a_1, a_2, a_3]^T$ (note this vector represents a specific force with corrections for gravitational and centripetal accelerations [1]).

The parameter measurement equation can be written as

$$z_k = [Hx_k + \zeta_k] \tag{4.12}$$

where H is the barometric height measurement vector and can be written as

$$H = \begin{matrix} 0_{8\times1}, & I_{8\times8} \\ 0_{1\times8}, & I \end{matrix} \tag{4.13}$$

where

$$z = [-\gamma \; v_{GPS} \; r_{GPS} \; h]^T \tag{4.14}$$

where

γ represents the vector of two parameters of orientation from the magnetometer sensor

v_{GPS} and r_{GPS} are the GPS estimates of velocity and position parameters

h is the estimate of the height from the barometric altimeter

H is the measurement matrix

x_k is a state vector

ζ_k error measurement vectors

SINS correction tasks can be solved using the following vector equation:

$$x_k = [x_k + K_k(z_k - Hx_k)] \tag{4.15}$$

The Kalman filter equations will generate the optimal estimate of the state vector x_k. It is critically important to understand that the matrices ϕ_k and H_K do not constitute fully observable pairs. Therefore, it requires very accurate calculations. In such tasks, algorithms for the computation of Cholesky factors of corresponding covariance matrices are generally used [1].

Development and Experimental Evaluation of Prototype UAV Navigation System

Development and experimental evaluation of the prototype UAV navigation system were successfully carried out by research scientists at the National Aviation University in Ukraine over the 2013–2014 duration. A set of sensors designed and developed at the National Aviation University were used in the evaluation of the prototype UAV navigation system. Several experiments including the laboratory static tests and various ground tests and flight tests were carried out for the realistic and experimental evaluation of the UAV navigation system.

Estimated linear velocities and coordinates as a function of time were obtained using equations. Measurements from the GNSS receiver and linear velocities are essential to determine the essential

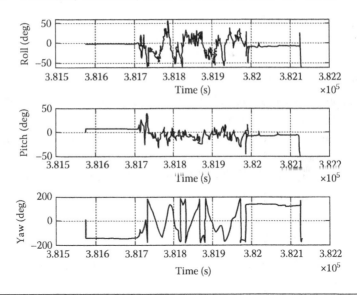

Figure 4.2 Graphic representation of velocities in northeast direction.

parameters. SINS errors can be used in plotting various parameters as a function of time. The calculated values shown are fairly in agreement with the simulation results obtained in the lab and coincide in the normal mode of the integrated prototype navigation system functioning. In summary, the measurements are of good quality and are fairly accurate. Time-dependent estimated values of roll, pitch, and yaw angles as a function of time can be seen in Figure 4.2.

SINS Correction Algorithm

Kalman filter equations play a critical role in error correction in the SINS. Equations generate vector of optimum estimate of $\mathbf{x}_k$ vector, which generates matrices $\mathbf{Q}_k$ and $\mathbf{R}_k$ that are covariance matrices of propagation errors in the SINS. Several experiments have been conducted by the author V. Kharchenko and his colleagues [1] to estimate and calculate current velocities, positions, and orientation parameters. These data are transferred to the AFCS, where the flight trajectory data are compared. Time dependency of estimated linear velocities and coordinates in the NED are represented in Figures 4.3 and 4.4, respectively. Note that the black line depicts the measurements from the GNSS receiver, while the gray solid line

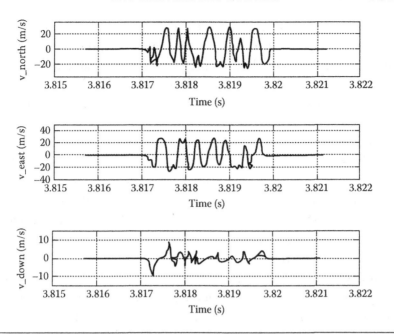

Figure 4.3   Graphic representation of velocity components in northeast down direction.

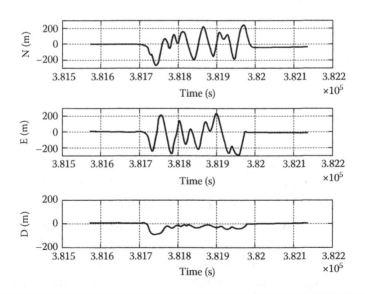

Figure 4.4   Graphic representation of the coordinates in northeast direction.

indicates the linear velocities estimated by the SINS. It is evident from these figures that these values coincide in the normal mode of INS functioning, when all measurements available are of good quality. Note that the time dependency of estimated roll, pitch, and yaw angles is shown in Figure 4.2. It is evident from these figures that INS algorithms work very efficiently even with the presence of noise and some error components at the measurements of various inertial sensors. Note that such a claim is true only if the GNSS measurement data are constantly available when needed. It has been observed that some velocity and position data will degrade after 10–20 s of INS autonomous mode of functioning most likely due to the presence of other error components in the MEMS inertial sensor measurement data.

Because of the wide application of future UAV systems, it is critically important to design and develop the most reliable and accurate navigation and AFCSs. This particular section identifies some prominent problems with their development and practical implementation. The prototype INS described here has been designed, developed, and evaluated by the scientists at the Ukraine Aerospace Center of National Aviation University during 2012–2013. In summary, it can be stated that the measured data confirm the efficient functioning and experimental validations of the UAV integrated navigation system. The author hopes that future experimental research efforts will focus on reduction of weight and size of the system, effective implementation of new error-free algorithms, including adaptive nonlinear filtering algorithms, and tight integration schemes.

Requirements of UAV's Automatic Flight Control System (AFCS)

This particular section identifies the critical requirements for the AFCS for the UAV platform with major emphasis on stabilization mode and navigation system accuracy. In future UCAVs, the design of the AFCS [2] must provide high reliability, accuracy, and vehicle platform safety under combat environments. In addition, the AFCS design must focus on the effectiveness of UCAV deployment in satisfying the conflicting mission requirements for robustness and high quality. A robust AFCS can significantly improve the flight control of the costly and complex UAVs, which have some uncertainties in its mathematical modeling and external disturbances due to aerodynamic conditions.

Critical Functions of AFCS [2]

The AFCS must be designed to provide smooth and stable flight performance to the UAV irrespective of weather conditions. Basically, the AFCS is designed to perform automatic control of angular stabilization and control of angular position of the UAV, trajectory control, and control of UAV platform during all flight phases from takeoff to landing. In brief, the AFCS should be capable of solving the following distinct responsibilities:

- Handing of manual control mode
- Automatic stabilization of the UAV angular position along three axes
- Automatic en route flight
- Automatic takeoff and landing
- Manual control mode operation
- "Scanning area" mode of operation
- UAV patrol mode of operation around specific point with the defined coordinates
- Improvement in the stability and control of the UAV flight characteristics
- Increase in the flight and maneuverability behavior of the UAV
- Restriction of dangerous conditions of UAV flight
- Activation of UAV horizontal flight mode of operation
- Automatic stabilization of UZV altitude
- Automatic course control of UAV
- Automatic stabilization and control of UAV true speed based on engine throttle control
- Automatic control of UAV I lateral plane (stabilization of the course line)
- Operative modification of the current flight route and flight mode of UAV by a command signal from the UAV operator
- Transmission of the telemetry information about basic flight and UAV navigation parameters through a radio channel to the ground control station (GCS)

- Control of raw, pitch, and roll angles according to the specified values received from the GCS
- Automatic preflight and flight control with indication of a failed stabilization mode
- Loud alarm when the UAV reaches deviations in the lateral and longitudinal movements

Critical Functions of the AFCS [3]

Comprehensive studies performed on the critical functions of the AFCS indicate that flight safety, angular stabilization of the UAV, and trajectory control are the essential functions of the AFCS, which can be briefly summarized as follows:

- Automatic control of the angular stabilization of the vehicle
- Control of the angular positions regardless of weather conditions
- Trajectory control for the safety of the vehicle
- Combined control of the UAV during all flight phases ranging from takeoff to landing

Principal Design Objective of the AFCS

The principal design objectives of the AFCS are to facilitate the work of the UAV pilot or operator, improve the safety of the vehicle, and increase the efficiency of the UAV. The AFCS must be designed to perform the automatic control of angular stabilization and of the angular position of the aircraft and absolute control of trajectory as well as of the UAV during all flight phases ranging from takeoff to landing.

Definitions of Operating Modes and Functions Associated with Modes

Manual Control Mode: This mode allows the transmission of electrical command signals from the receiver directly to the actuators associated with the UAV. These actuators provide appropriate control of roll, pitch, and yaw angles necessary for the flight control of the aircraft.

Automatic Angular Stabilization of the UAV Aircraft [4]: During this mode, automatic angular stabilization along the three main axes is automatically provided. When the UAV operator releases the handles at the transmitter, control of roll, pitch, and yaw angles should be carried out by the AFCS at the earliest possible time to preserve the safety and flight integrity of the UAV. At this moment, the UAV should automatically switch to horizontal flight mode. In addition, engine throttles should be controlled directly from the transmitter or automatically from the AFCS.

Return to Home Mode: This particular mode of operation requires that the UAV aircraft automatically return to home base or to the starting point at the specified altitude or height. Manual assistance is available if needed.

Automatic En Route Flight: During this mode of operation, automatic UAV flight with previously programmed waypoints could continue with manual assistance from the UAV operator. Programming the flight plans should be performed by the methods specified as follows:

- Manual inputs of waypoints defined by geographical coordinates
- Automatic upload from an external source of previously planned flight route
- Change of the flight plan
- Skipping one or more waypoints
- Setting the turn lead distance or flight with passage of waypoint
- Quick transition to the new flight plan with ability to return to any point of the original route
- Flight to perform an emergency landing
- Flight on the selected navigation waypoint or promptly selected waypoint
- Flight on parallel trajectories with displacement on a selected distance up to 1 km related to the intended trajectory
- Addition of new waypoint to existing trajectory plan

It is necessary to emphasize that when one intercepts a UAV automatic flight plan en route, it should continue to perform its specified route following the next waypoint, even if all waypoints are already passed. Note that the UAV has to return to home base or starting mode of operation [3].

Automatic Takeoff Mode of Operation: This particular mode of operation involves preparation to taxiing for a specific duration, acceleration, takeoff, and initial climb at a specified speed, which continues to increase until the aircraft reaches a level flight. These operations should be completed in the automatic takeoff mode.

Automatic Landing Mode: Automatic landing mode involves appro priate stages of planning, altitude leveling, altitude holding, parachute deployment if required, and keeping the aircraft running after a safe landing on the landing strip.

Essential Components or Subsystems of AFCS

Essential components or subsystems of the AFCS can be summarized as follows:

- High-speed computer with built-in programmable logic controller
- Input–output modules
- Control panel located inside the GCS

The control panel should allow the operator to select system operating modes such as on, off, automatic takeoff, and automatic landing and obtain information regarding the state of electrical switches, push buttons, and UAV flight parameters such as speed and altitude. The panel should clearly indicate various functions of the control panel elements, which should be controlled by the computer.

The AFCS should clearly mark the three distinct automatic control subsystems and control channels such as engine throttle channel, longitudinal channel, and lateral channel. Note that the longitudinal control channel of AFCS should be realized through the elevator surface, while the lateral channel through ailerons and rudder surfaces.

Importance of Roll and Yaw Channels: The roll and yaw channels form a lateral channel of the AFCS. Note that the yaw channel plays a critical role in providing the right amount of yaw damping and in eliminating the slip when performing coordinating turns in automatic mode of lateral channel.

Important Function of Longitudinal Channel: The longitudinal channel of the AFCS, in the automatic mode of operation, should stabilize the pitch angle and should provide the desired trajectory of the motion of the UAV aircraft in a vertical plane.

Executive Roles of Longitudinal and Lateral Channels: Note that the executive functions of the longitudinal channels of AFCS are worth mentioning. These channels provide mechanical connections through cable and thrust rocking, traction in the channel, and automatic traction control in the engine. It should be stressed that control commands should be generated in the form of standard AFCS pulse-width modulated signals, which are considered most suitable for the major types of actuators [4].

Critical Functions of AFCS

The AFCS should provide the following critical functions:

- Receive necessary signals from the operating sensors such as navigation and air speed sensors
- Process information from the sensors abroad the UAV aircraft
- Generate control signals for the surface actuators
- Provide mode switching functions
- Provide flight and preflight control functions of serviceability of all levels of hierarchy of AFCS
- Provide its own input and output information and its delivery to the appropriate system

Software for AFCS

To meet specific requirements for configuration, adjustment, and testing of the AFCS, special software capable of operating on the Linux system needs to be created. This particular operating system is capable of generating different adaptive and robust schemes and algorithms of the UAV flight control in accordance with the tasks to be performed by the system [4].

The tasks should be capable of creating requirements for each UAV flight mode with its own scheme and control algorithm, which is best suited for this particular mode. It is important to make sure that the specialized software contains a set of basic blocks for the effective implementation of the control loops, including signal conversion components, proportional–integral–derivative controller, mathematical functions, logical blocks, and other functional loops. It is also necessary to create postflight analysis and preparation with emphasis to save all the data on the positions and actions of the UAV operator.

Properties of Specialized Software

Specialized software is designed for a specific system with most unique properties, which will ensure optimum performance, including robustness, reliability, and foolproof security. Properties of specialized software must include the following items:

- Optimum system performance with foolproof security
- Performance specification flexibility for particular parameters
- Optimum robustness and adaptability capability
- Ability of the software for interactive work with the AFCS if required

Basic Performance Specification Requirements for the AFCS Module

Basic performance specification requirements for the AFCS module are summarized in Table 4.1.

Accuracy Requirements for AFCS Parameters [4]: The AFCS module with other related systems should meet the automatic stabilization requirements of the UAV orientation relative to its center of gravity (meaning its mass) with the following admissible errors:

- Pitch angular error not exceeding ±0.5°
- Yaw angular error not exceeding ±0.5°
- Roll angular error not exceeding ±0.5°
- Automatic stabilization barometric altitude with admissible error at altitudes up to 3000 m not exceeding ±5 m
- Automatic stabilization and specified course control with admissible error not exceeding ±0.5°

Table 4.1 Typical Performance Specification
Requirements for the AFCS Module

AFCS MODULE SPECIFICATIONS	TYPICAL VALUES
Operating altitude range (ft)	TBD
Velocity range (ft/s)	TBD
Ambient temperature range (°C)	(−20 to +50)
Humidity (%)	TBD
Level of radiated interference (dB)	TBD
Resistance to EM radiation (dB)	TBD
Vibrations at the attaching points	TBD
Shock survivability	TBD
Peak acceleration (cm/s^2)	TBD
Airborne power supply voltage (max)	15 V
Power consumption (W)	3 (max)
Module dimensions (mm)	$100 \times 50 \times 15$ (max)
Weight (g)	75 (max)
Sealing protection specifications	IP54
Average service life (years)	50

Indication of Emergency Conditions from AFCS Algorithms

It is absolutely essential for the AFCS algorithms to provide an indication of possible emergency conditions. In other words, the AFCS algorithm architectures must be capable of providing the indication of possible emergency situations specifying specific sources of trouble. This requires that the state of AFCS "ACCIDENT" should include stop execution algorithms inside the AFCS module, flush the corresponding light-emitting diodes on the board of AFCS module, and simultaneously display alerts on the GCS monitor.

Programming and Adjustment of AFCS

Programming should be designed to select appropriate algorithms and to set the values of specified performance parameters such as transition coefficients, which determine its configuration and functioning.

During its functioning mode, changing the settings in accordance with the terms and purposes of the AFCS exploitation is possible. In addition, the values of the programming parameters must be saved in a nonvolatile memory card of the AFCS and must be stored when the input power is removed. It is necessary that in the AFCS operation, algorithms must be designed and developed to provide appropriate solutions for the most frequently encountered problems in the operation of the UAV flight control.

Procedures Necessary for AFCS Adjustment: The following are the appropriate procedures needed for adjustment in the AFCS for satisfactory operation:

- Check all the modes needed for stabilization of the AFCS.
- Check carefully the performance of the navigation, landing, and takeoff modes.
- Check all the control law coefficients widely used in the AFCS.
- Check the constraints on the deviations for all control surfaces and the engine thrust level.
- Check the values of specified control signals.
- Check all restrictions imposed on the control channels including engine thrust, longitudinal, and lateral channels.
- Check the activation of return to home mode while simulating failures of the sensors aboard.
- Check all electro-optical (EO) and electromagnetic (EM) sensor connections deployed in the navigation system, air speed sensor, and other electromechanical sensors.
- Check the operational modes of the airborne computer of the AFCS.

Concluding Remarks on AFCS: Research studies undertaken by the author on the UAV AFCS seem to draw some meaningful conclusions, which can be highlighted in this section. The studies seem to stress that during the design phase of AFCS for the UAV, significant importance must be given to the accuracy requirements. In addition, critical functions of the AFCS software must be defined explicitly with particular emphasis on the quality of the software and adjustment requirements, technical requirements for the AFCS board, its operating modes, and improvements of the functioning of UAV.

UAV Fault Detection and Isolation [5]

UAV designers and research scientists believe that the development of techniques to detect and isolate faults is of paramount importance to preserve the mechanical integrity of the UAV aircraft. Regardless of the system type, undetected faults in a system can have catastrophic effects, including human lives, financial losses and environmental pollution. Properly implemented fault detection and isolation (FDI) schemes can be most effective particularly in hospitals, nuclear power plants, submarines, UAVs, and manufacturing companies. To discuss the FDI problems, the following commonly accepted terms are widely deployed [5]:

- *Unknown inputs* such as system noise level, measurement errors, modeling errors, and system parameters variations.
- *Type of fault* such as critical performance parameter variation.
- *Residual fault* such as a fault indication based on deviation between the laboratory measurement and modeling-based estimate.
- *Fault detection* such as determination of the faults present in the system and their time of detection.
- *Fault isolation* clearly indicates the presence of the fault in the system.
- *Fault accommodation* confirms the presence of the fault in the system and the evidence of maintaining safe system operation.
- *Analytical redundancy* reveals two methods for determination of a variable, one method using an analytical analysis and the other method using a computer modeling.

FDI experts believe that FDI methods can be divided into two categories [5]:

1. One category deploys a plant model.
2. Other category does not make the use of a plant model.

Experts believe that the traditional FDI method strictly relies on redundant hardware or sensors. Nevertheless, this method is very costly despite its suitability and popularity. Another most popular approach is based on limit-value checking of the characteristic variables such as temperature and air pressure. The limit-value checking technique remains one of the most widely implemented FDI methods in the manufacturing industry because of its simplicity and minimum complexity. The limit-value checking technique is only reliable if the faults are significant and long lasting. This is true because the thresholds are set at high levels to avoid false alarms caused by the random system fluctuations. Furthermore, in closed loop systems, the control laws tend to dampen the effects of the faults, and therefore, simply checking the amplitude of the system output signals does not give a reliable picture of the overall system's health. Manufacturing experts think that the tolerance or threshold method is most reliable and offers quality control products. But this particular method is strictly dependent on many years of practical experience. Aerospace engineers believe that the limit-value checking method is generally used for engine health monitoring. Nevertheless, advanced failure detection techniques must be deployed.

FDI scientists feel that frequency analysis of the measured signals can provide invaluable information on the machine or system's health. This method is particularly most popular if the faults cause an increase in machine vibrations of higher amplitudes. The frequency spectrum of these vibrations can be used for FDI determination. A technical paper or report discussing the vibration-based FDI methods can be a good source of FDI information.

Research studies performed by the author indicate that the use of "expert systems" also known as "knowledge-based methods" is the most popular FDI approach. Note that the experts rely on measurable symptoms, machine performance history, and other performance parameters to detect and isolate faults. Note that the fault detection is based on qualitative information, which can be readily available from the system's health history, maintenance records, and human observations including smell or unusual sound.

Fault isolation can be based on "if-then" logic or pattern classification techniques using neural networks (NNs). Appropriate technical papers must be reviewed for alternate techniques on the subject concerned. It is imperative to mention that expert systems have received considerable attention over the last two decades or so.

Feedback control theory has brought powerful techniques in mathematical modeling, which has been made feasible by the rapid progress of modern computer technology. This technology has allowed research scientists to deploy direct replacement of redundant hardware or redundant sensors. FDI engineers believe that a model-based

FDI system is the leading control system theory, which could play vital roles in solving the FDI problems. Note that the models must be robust to modeling errors when applied to a real system. Note that an inadequate control law can result in serious instability in a real system while an inadequate FDI scheme can result in high false alarm rates and undetected faults. However, the combination of an FDI scheme and a control system known as a fault tolerance control system (FTCS) is essential. Thus, this FTCS can benefit from the ability to compensate for faults detected and isolated by the FDI scheme while maintaining satisfactory system performance. Experienced engineers noticed that the physical redundancy and traditional limit-value checking techniques can provide remarkable results compared to the use of complex control theory and computer simulation techniques. In the case of complex aerospace systems where production costs are generally high, the use of redundant sensors may not be significant compared to efforts required in modeling of such a complex and costly system. Complex system designers feel that there are some applications where sensor redundancy may not be an option, and therefore in such a case, a model-based scheme becomes a viable approach. This is particularly true for UAVs where limited onboard space, weight restriction, and low-cost are the principal procurement requirements.

Model-Based FDI Approach: A model-based FDI approach can be very suitable for some applications. For example, a UAV aircraft can be divided into three distinct subsystems, namely, surface actuators, aircraft structure, and EO and EM sensors. The actuators control the surfaces such as elevators, ailerons, and rudders; the process could include the UAV airframe; and the EO and EM sensors will be recording instruments aboard the UAV. The model-based schemes involve two stages, namely, residual generation and residual evaluation. Furthermore, a system model is used to generate a residual, which is generally a function of the difference between the model estimate and the real measurement. This particular stage is known as residual generation. However, the FDI decision is made in the residual evaluation stage. In summary, it can be stated that in most cases the method of residual evaluation is strictly dependent on the method of residual generation.

Faults in UAV Manufacturing Facility [5]: In such a manufacturing facility, the faults can be divided into three distinct categories as mentioned in the following:

1. Actuator faults
2. Process faults
3. Sensor faults

Specific Details of the Fault in UAV Manufacturing Facility: Any actuator fault can cause a fault in the control surface of the UAV platform or aircraft. Such faults are considered additive faults in the sense that they influence the system with an additive term. Note that sensor faults are also considered additive faults but they influence the instrumentation parameters of the system or UAV aircraft. These sensors could include sensor biasing parameters, sensor drifts, and sensor failures. Process faults can be either additive or multiplicative. Parametric faults are considered multiplicative process faults leading to the fact that such faults influence the manufacturing plant output by the product of another variable, which results in changes in the plant parameters. Any deterioration of the UAV airframe will be considered a parametric fault. Note that process faults can be characterized as additive faults. In some cases, it is important to classify different faults in different categories. Furthermore, different FDI schemes are better suited for different types of faults. For example, parameter estimation methods are best suited for parameter faults, while the observer-based schemes are most suitable to detect actuator and sensor faults. In addition to different fault categories, faults can be quick varying (abrupt) or slow varying (incipient). In other words, a quick-varying fault would create sudden system performance change, while a slow-varying fault would see a drifting-type effect.

Various Model-Based FDI Methods [6]: Close examination of technical published papers [5] reveals that a wide variety of model-based FDI methods are available:

- Observer-based method
- Parity-based method
- Fault detection–filter based method
- Parameter estimation–based method

- NN-based method
- Eigen structure–based method

Observer-Based Method for FDI: In this particular method, fault detection is based on the inputs provided by the observer. However, it depends whether the observer reports the faults at regular intervals or randomly. Note that for reliable information, reporting the faults at regular intervals will be necessary.

Parity Space–Based Method for FDI: Published technical papers on FDI methods indicate that the parity space–based method for model-based FDI schemes is most popular. In this approach, a number of plant observations are sampled from previous time instants to current time instant. The residual generated is calculated as a function of these sample measurements and a user-defined matrix. Designing the matrix in an appropriate way will set the residual to zero when no faults are present and nonzero will appear when faults are present. In brief, the residual is zero when there are no faults. This method appears to be very simple as well as reliable.

Fault Detection Filter–Based Method: In this method, the fault detection is strictly dependent on the filter output or the residual evaluation. However, the residual evaluation is a function of filter response.

Parameter Estimation–Based Method: This method is based on the estimated values of critical system parameters. This requires priority on system performance parameters based on system specification-based parameter values such as radar transmitter efficiency greater than 75%, mean time between failures exceeding 5000 h, and minimum weight of 60 lb. Here, the critical performance parameters for the airborne radar transmitter include the efficiency, mean time between failures, and the minimum weight are 75%, 5000 h, and 60 lb, respectively.

Neural Network–Based Method: Artificial NNs consist of a large number of simple processing elements known as neurons, which are interconnected together via channels known as connections. The structure formed is inspired by the structure of the brain's biological nervous system. Note that due to highly interconnected networks, it could be difficult to visualize and predict the performance of NNs.

It is therefore suggested that the NN model selected be thoroughly tested before its implementation.

One of the major benefits of NN technology is its highly interconnected structure, which will make it fault tolerant. In other words, the system performance is not significantly degraded if one of the links or neurons is faulty. For example, if thousands of thousands of inputs are connected to one neuron, then the output of this neuron will be most robust to faults in one or more of its inputs. The main property of NNs makes them superior than traditional modeling methods in their online adaptive capabilities.

Using an appropriate training algorithm, an NN can upgrade its own structure in real time, which will better suit the input/output data. As a matter of fact, an NN trained to operate in a specific environment can be easily retrained to deal with minor changes under the operating environments. This property is particularly useful when modeling the time-varying systems.

Because of their highly interconnected structure, it will be extremely difficult to predict the performance of NNs. Furthermore, due to their nonparametric modeling approach, the NN approach is best suited for applications where theory is poor but the training data are plentiful. NN concepts can be used for a variety of applications ranging from traffic monitoring to predicting the price of gold. Using input from humans and webcams, the NN-based circuits compute important traffic parameters such as traffic flow, traffic density, traffic flow, and automobile speed. NN approaches have been used in the field of finance because of data speed and accuracy. Using previous gold prices and Dow Jones indices, the NN can be designed to accurately predict gold prices.

Each NN layer is considered a multilayer perception (MLP) and real gold price data are used to train and test the NNs. Note that an MLP NN with 3-3-1 structure trained with the conventional algorithm also is used to model the solar activity of the sun. This model receives information regarding the solar magnetic field properties.

Studies performed by the author seem to indicate that the most common problem with NN technology is that of overtraining the NN. Furthermore, estimations are accurate during the training process but are poor when exposed to new data. In other words, the generalization capabilities are poor. It is interesting to point out that NN-based training can be stopped when the test set error starts to increase. Note that the offline training stopping criteria selected are based on checking NN convergence as well as avoiding the overfitting phenomenon. The data sets provide the following procedures:

- The training data set is used to train the NN with learning when the switch is ON.
- The training data set is used to train the NN with learning when the switch is OFF.

Note that the NN offline training is stopped based on the following two criteria:

Criterion 1: The root mean square (RMS) error of the test data set increases for more than 100 consecutive times.

Criterion 2: The rate of change of RMS per epoch is less than 0.1% for more than 100 consecutive epochs for both the test data set and training data set.

One epoch represents one pass or one event through the whole data set. Note criterion 1 is the overfitting criterion whereas criterion 2 checks if the NN is converged.

Comments on NNs: In this subsection, the author has presented the NN structure for possible application in the sensor FDI accommodation scheme and the flush air data sensor (FADS) system. The NN technology is selected due to its excellent generalization capabilities, compact structure, and fast execution times. One of the principal reasons for selecting the NN model is its online adaptive capabilities in comparison to traditional methods, which rely on a fixed mathematical model such as Kalman filters. Later in this chapter, the performance of the NN model in comparison to Kalman filter will be summarized with emphasis on adaptive processing efficiency. The method to train NN models has been discussed in great detail. It has been noticed that the online training in the FADS system is not possible due to the requirements of additional instrumentation. The NN training stopping criteria are defined based on the need for checking the NN convergence and avoidance of overfitting the NN structure.

Kalman Filtering [7]

Major improvement in an INS in cooperation with Global Positioning System (GPS) strictly depends on the prediction and evaluation of INS errors. Results of mathematical modeling will confirm the feasibility of this approach for improving the accuracy of an INS in offline mode.

Research studies performed by UAV design engineers indicate there is a serious problem of weight and size characteristics of airborne equipment. Therefore, for UAVs it is most appropriate to use the integrated INS/GPS system as a central navigational unit. Note the core of the INS/GPS complex system is called a SINS, implemented as a MEMS, and integrated with the GPS or any other space-based Satellite Navigation System (SNS).

In the existing complex INS/GPS system, evaluation of the parameters that characterize the state vector of the UAV vehicle can be performed by extended Kalman filter (EKF) technology [7].

Note that performance parameters can be used for adjustment of SINS performance. Despite its usefulness for system integration using Kalman filtering, it requires a significant control processing time to perform the calculations involved. In addition, there are a number of problems with the onboard implementation of the Kalman filter. The most important problem with this technique is the divergence phenomenon. Note that the main sources of errors in the SINS are the drifts of the gyroscopic sensors and accelerometers, which are due to nonstationary matrix of transition from one coordinate system associated with the UAV platform to another navigational unit due to nonstationary stochastic processes. This generates additional difficulties in identifying additional SINS errors by the optimum filtering methods.

To avoid the previously mentioned problems, a series of modifications in the Kalman filter structure were implemented including the robust and adaptive filtering algorithms and other high-performance algorithms. Furthermore, to reduce the central processing unit time, the modified and compact Kalman filter is finally used. In addition, unobservable parameters of the state vector besides the parameters of the angular orientation are corrected from additional sources of information using magnetometers, accelerometers, and other sensors capable of measuring these parameters. Other methods are available, which use integration of different systems, and they work well in actual practice, particularly in the Doppler inertial navigation systems. Many research studies were conducted to improve the algorithms of integration of SNS and INS systems, and significant improvement was achieved and observed in actual practice.

The principal objective of the improvement program initiated by the research scientists at the Aviation Computer-Integration Department at Kiev (Ukraine) was to demonstrate the benefits of fusion of navigation information on the UAV board not on the basis of the reduced Kalman filter structure but through the deployment of the compensation scheme. Note successful implementation of the Kalman filter requires an extrapolation of errors in the navigation system being corrected, which allows, in the case of INS malfunctions, to maintain accuracy requirements of INS system at the right level at least for some time. Note that the problem is stated at the time of development of algorithms for extrapolation of SINS errors in the compensation circuit. The compensation circuit provides accurate, autonomous functioning of the INS system when a GPS signal is not readily available.

Prediction and Solution for INS System Errors: Both the compensation scheme and the algorithm for information processing use the method of compensation, which is relatively simple compared to Kalman filtering. The evaluation navigation parameter X can be written as

$$X = [X_{SINS} - F(p)(X_{SINS} - X_{SNS})] \qquad (4.16)$$

where

X_{SINS} and X_{SNS} are the navigation position and velocity component parameters, respectively

X is the evaluation navigation parameter

$F(p)$ is the transfer function of the compensation filter

For the compensation circuit, it is necessary to synthesize a dynamic filter. Synthesizing such a filter can use the procedure that deploys the regularization technique that has the following expression for the filter transfer function:

$$F(p) = [(3T_p + 1)]/[(T_p + 1)(T_p + 1)(T_p + 1)] \qquad (4.17)$$

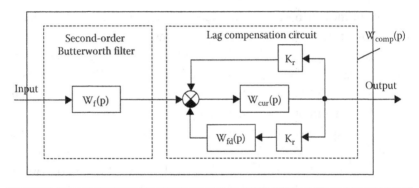

Figure 4.5 Critical elements of the compensation circuit filter.

The compensation scheme as illustrated in Figure 4.5 and the algorithm of the information processing, which uses the method of compensation, have a relatively simple form compared to the optimum Kalman filtering. It is equally important to mention that for the compensation circuit shown in Figure 4.5, a dynamic filter has been synthesized using the procedure that uses the regularization method. The structure of the synthesized filter present in the compensation circuit is clearly shown in Figure 4.5.

Various feedback transfer functions in the compensation circuit and integrated processing circuit can be defined as follows:

$W_{fd}(p) = W_f(p)$ is the feedback transfer function of the lag compensation network or circuit as shown in Figure 4.5

K_r = regularization parameter

$W_{corr}(p)$ = transfer function of the synthesized correction block

$W_{comp}(p)$ = transfer function of the lag compensation network or circuit

δ_1 = low-frequency error component of SINS

δ_2 = high-frequency error component of the SNS

It is necessary to mention the frequency characteristics of interferences δ_1 and δ_2 that differ significantly, which occur in the INS/GPS system [7].

F(p), as shown in Figure 4.5, is the output of the compensation filter [8].

For complete error analysis, one has to assume the structure of a mathematical model of SNS error and keep a track of current error of compensation scheme at the output of the compensation filter F(p); it is possible to use parametric identification to correct the mathematical model in such a way that it recreates a time-varying current error of SINS as accurately as possible. A mathematical model can utilize SINS error models in which the Kalman filter algorithms are based. In other words, it is important to remember that Kalman filtering [7] is strictly based on the SINS errors. It is equally important to mention that the parameters of the angular orientation can be corrected from the information supplied by the magnetometers and accelerators.

Using these mathematical models, equations for acceleration errors, velocity errors, variations in earth's radius, altitude errors, and UAV angular velocity errors can be written as follows:

Starting from the acceleration errors, one can write the following expressions:

$$[\Delta dV_E/dt] = [A + B - C + D + E - (F_1)(F_2) + G] \qquad (4.18)$$

where

$$A = [a_{N\,\alpha}H - a_H\,\alpha_N] \qquad (4.19)$$

$$B = [\Sigma^3\,b_{1i}\,\Delta a_i] \qquad (4.20)$$

$$C = [\Delta V_H\,U(t)\,\cos(\varphi)] \qquad (4.21)$$

$$D = [\Delta V_N\,U(t)\,\sin(\varphi)] \qquad (4.22)$$

$$E = [\Delta R_N/R_E(U(t)(V_H\sin(\varphi) + V_N\cos(\varphi))] \qquad (4.23)$$

$$F = [(\Delta V_E/R\,\cos(\varphi)) + (V_E\,\sin(\varphi)\Delta R_N/R\,\cos^2\varphi\,R_E)$$
$$\times(V_H\cos(\varphi) - V_N\sin(\varphi))] \qquad (4.24)$$

$$G = [(\Delta HV_E/R^2)(V_H - V_N\,t\,g\,\varphi)] \qquad (4.25)$$

$$\text{Note } (F1)\,(F2) = F \text{ and } F^* = (F^*1) \times (F^*2) \qquad (4.26)$$

where symbol * indicates transpose matrix.

$$[\Delta dV_N/dt] = [A^* + B^* - C^* - D^* - E^* - (F_1^*)(F_2^*) \qquad (4.27)$$

where

$$A^* = [-a_E\alpha_H + a_H\alpha_E] \qquad (4.28)$$

$$B^* = [\Sigma^3\,b_{2,I}\,\Delta a_i] \qquad (4.29)$$

$$C^* = [\Delta V_E \, U(t) \sin(\varphi) - \Delta V_H \, (d\varphi/dt)] \tag{4.30}$$

$$D^* = [(\Delta R_N/R_E) \, V_E \, U(t) \cos(\varphi)] \tag{4.31}$$

$$E^* = [(\Delta V_N/R)V_H] \tag{4.32}$$

$$F^* = [\Delta V_E/R \cos(\varphi) + V_E \sin(\varphi)\Delta R_N/R \cos^2 \varphi \, R_E]$$
$$\times [V_E \sin(\varphi)+(\Delta H/R^2)(V_E{}^2 \, tg\varphi + V_N V_H)] \tag{4.33}$$

$$[\Delta V_H] = [a_E \alpha_N - a_{N\alpha F} + \Sigma^3 \, b_{3,i}\Delta a_i + \Delta V_D U(t) \cos(\varphi)$$
$$+ \Delta V_{N\varphi}(t)] - [(\Delta R_N/R_E)U(t) \, V_E \sin(\varphi) +(\Delta V_N/R)V_N]$$
$$+ [(\Delta V_E/R \cos(\varphi) + (V_E \sin (\varphi)\Delta R_N/R \cos^2 \varphi \, R_E)]$$
$$\times [V_E \cos \varphi] + g_e \, [-(2\Delta H/a) +1.5e^2 \sin\varphi \cos \varphi \Delta R_N/R_E]$$
$$- [(\Delta H/R^2) \, (V_E{}^2 + R_N{}^2)] \tag{4.34}$$

$$[dE/dt] = [(\Delta V_E(t)(R_E/R \cos \varphi(t))] + [(\Delta R_N(t)) \, (V_E \sin \varphi(t)/$$
$$R_E R \cos^2 \varphi(t))] - [\Delta H(t) \, R_E \, V_E(t)/R^2 \cos \varphi(t)] \tag{4.35}$$

$$[\Delta d \, R_N/dt] = [\Delta V_N(t)(R_E/R)] - [\Delta H(t) \, R_E \, V_N(t)/R^2] \tag{4.36}$$

$$[\Delta d \, H(t)/dt] = [\Delta V_H(t)] \tag{4.37}$$

Definitions of various symbols used in the earlier equations can be interpreted as follows:

R_E = the radius of earth = radius of the earth sphere
R = current UAV altitude = $R_E + H$
$U(t) = [2\omega_E + d\lambda(t)/dt)]$
$\omega_E(t)$ = angular velocity of earth's rotation

Description of Various Errors

1. Calculation errors of reduced coordinates of UAV:

$$\Delta R_E(t) = [\Delta\lambda(t)R_3]$$

$$\Delta R_N(t) = [\Delta\varphi(t)R_3]$$

2. Calculated errors of the geographic coordinates:

$$[\Delta\lambda(t)], \quad [\Delta\varphi(t)], \quad \text{and} \quad [\Delta H(t)]$$

3. Calculated errors of the projections of earth velocities:

$$[\Delta V_E(t)], \quad [\Delta V_N(t)], \quad \text{and} \quad [\Delta V_H(t)]$$

4. Current altitude R can be written as

$$R = R_E + H$$

where

R_E is the earth radius

H is the UAV operating height

5. SINS accelerometers errors can be written as

$$\Delta a_i$$

where i varies as 1, 2, 3.

6. Current values of the projections of the apparent acceleration on the axis of the INS can be written as

$$a_H, a_E, a_N,$$

which are defined as the current values of the errors in the plane elevation (E) and vertical azimuth plane, which are dependent on the SINS angular velocity sensor errors. Using these errors, one can generate mathematical models using Kalman filter technology or using the hypothetical mathematical models of SINS sensor errors [7]. According to modeling experts, using hypothetical models of errors can be estimated for the SINS sensor errors.

The statistical analysis of the SINS errors must be performed in order to formalize the structure of the hypothetical mathematical models for SINS errors. Note the UAV dynamic parameters must be considered in determining the earth speed projection values such as $\Delta V_E(t)$ and $\Delta V_N(t)$. After that one can undertake statistical analysis for changes in the errors of SINS sensor. The statistical analysis can be performed using mathematical modeling with different initial conditions that can have impact on error behavior. Essentially, they are deterministic components of errors of the primary information sensors such as accelerators and angular rate sensors, UAV vehicle flight direction, and the latitude of the UAV location.

Calculation of Estimated Error of UAV Speed in SINS Algorithms

Estimated UAV errors can be calculated from the SINS algorithms. It is also important to remember that the velocity errors have sinusoidal dependence in the amplitude response (A), which is strictly the function of acceleration errors and has an oscillation frequency given by the following equation:

$$\omega_o = [a/g_H(1 - H/a - 0.5\ e^2 \sin^2 B)]^{0.5}, \qquad (4.38)$$

where

ω_o is the angular oscillation frequency
$a = 6.378388 \times 10^6$
H is the UAV flight altitude
B is the geographical latitude of the UAV at the given location
e is a constant and has a value of 0.820365772

$$g_H = [-g(1.5 + 5.2884 \sin^2 B)]\ [1 - 2H/a\ (1 - e \sin^2 B)] \qquad (4.39)$$

Inserting the given values in Equation 4.39, one can compute the magnitude of parameter g_H. Hypothetical mathematical models of calculation of errors for the latitude and longitude can be written as follows:

$$[\Delta(d\lambda/dt)] = [\Delta V_E/(R_2 + H)\cos B] \qquad (4.40)$$

$$[\Delta\varphi] = [\Delta V_N/R_1 + H] \qquad (4.41)$$

It should be noted that Equation 4.40 is valid for the calculated errors in latitude, while Equation 4.41 is valid for errors calculated for longitude. In these two equations, the parameters R_1 and R_2 represent the operating ranges, H indicates the UAV flight altitude, and ΔV_E and ΔV_N are the errors calculated for the velocity component in the directions mentioned. Note that the previously mentioned parameters can be obtained from the kinetic equations of SIMS sensor modeling.

Highlights of the computer analysis can be summarized by the following comments:

- The information provided by the error sensors seems to indicate that a phase shift ϕ of sinusoidal nature is appearing in the calculation errors of angular velocity components.
- There is also an additional component that depends on the current heading in $\Delta V_E(t)$ and $\Delta V_N(t)$ velocity components.
- During system integration implementation compensation schemes, it is possible to monitor the calculated errors, in addition to monitoring the error calculations for various coordinates.

- Note that the parameters shown in Equations 4.39 and 4.40 can be obtained directly from the algorithms for the solution of the SINS kinetic equations [7].
- The SINS error model can be deployed as a linear aggression with other parametric identifications of the coefficients of the SINS model.
- If it is necessary, adjustment of the model parameters can be carried out at appropriate times.
- Measure of the deviation of the SINS errors can be obtained from the error model in the compensation circuit.
- The analysis of the signals after the filtered parameters indicates the presence of random variations, because the dynamic filter with function $F(p)$ in the compensation circuit shown (Figure 4.5) cannot completely eliminate the high-frequency SNS error. Under these circumstances, the tasks of averaging and filtering of the estimated parameters remain pending.
- The following criteria for the proximation or approximation can be applied for maximum deviation, mean deviation, and RMS deviation.
- Note that parametric identification of the model used can be carried out using the gradient identification algorithm. Gradient identification algorithm can be materialized as described in the following paragraph.

Assume some initial value of the vector of parameters of the error model deployed. Then determine the residual value by solving the equations involved. By perturbing the parameters of the model errors by a certain amount, one can find the corresponding values of the residuals. An iterative process can be used to improve the estimates of the model parameters, provided the residual function decreases. Various modification gradient procedures are available. The results of the error modeling can be seen in Figure 4.6 based on the steepest decent method.

- The approach described here requires the integration of both the INS and the SNS based on the compensation circuit, which is fast and uncritical to nonstationary random processes. However, the outcome of the approach is strictly dependent on the compensation circuit as illustrated in Figure 4.5.

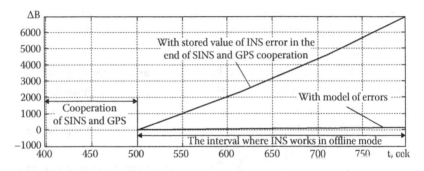

Figure 4.6 Simulation results for Strapdown Inertial Navigation System and Global Positioning System with the compensation circuit being active just for 500 s duration.

- Note that the method of predicting the changes of the SINS errors strictly depends on the design of the compensation circuit and also on the procedure of identification used for the parameters of hypothetical models of SINS sensor error at the junction of SINS/GPS incorporation. Note that the SINS/GPS combination can significantly decrease the SINS errors, while it functions in the offline operating mode. Space scientists and UAV project managers feel that the gradient method of identification is potentially the best tool for space systems.

- Note the algorithms recommended for the prediction of SINS errors were evaluated using mathematical modeling. This modeling can be performed using the visual simulation software Simulink®, which is the most critical component of the computing system MATLAB®.

- Identification of the parameters of a hypothetical model can be accomplished by the steepest descent method. However, for averaging and filtering of the estimated parameters, the least-squares method will be most accurate.

- The results of modeling the joint SINS/GPS systems with the compensation circuit operating for the first 500 s can be seen in Figure 4.6. Over this duration, search of parameters of a hypothetical model can be conducted. After 500 s, SINS is turned into offline mode. In this particular mode, the calculation errors of the latitude can be compared while taking into account the SINS errors that are stored at the end point of its functioning along with GPS or with extrapolating of its errors. However, in the latter option, the accuracy in the offline mode can be significantly improved. It will be interesting to note that Figure 4.6 clearly indicates the stored values of INS error in the end of SINS/GPS system corporation [5].

Role of Compensation Circuit Filter in the Joint SINS/SNS System Operation

Critical elements of the compensation circuit filter are shown in Figure 4.7. Close examination of this compensation filter indicates that SINS is getting velocity and position correction signals for the mixture, vertical channel correction signals from appropriate source and parameter correction, and corrected parameters of angular orientation signals from an appropriate source. The output of the SINS sensor is fed to SNS. The outputs from SINS and SNS are fed to the mixer, whose output acts like an input to the compensation circuit filter as shown in Figure 4.5. F(p) is the transfer function of the compensation circuit filter. Block diagram shown in Figure 4.7 provides complete information of integrated processing using computer techniques.

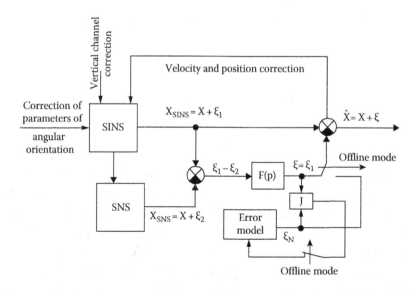

Figure 4.7 Block diagram of the integrated processing network using the compensation technique.

Parameter $W_{fd}(p)$ is equal to $W_f(p)$ where the subscript fd stands for feedback and $W_f(p)$ is the feedback transfer function of the "lag" compensation circuit [5]. Other circuit parameters are defined as follows:

- K_r is the regularization parameter.
- $W_{cor}(p)$ is the transfer function of the synthesized correction block shown in Figure 4.7.
- $W_{comp}(p)$ is the transfer function of the lag compensation circuit.
- ζ_1 represents the low-frequency interference signal or low-frequency error with minimum distortion and damps the interference ζ_2, which represents the high-frequency error of the SNS. This particular parameter minimizes the integrated system error. In the case of frequency characteristics of interference ζ_1, parameter ζ_2 differs significantly, which takes place in the INS/GPS system. Under these circumstances, the output of the filter of transfer function $F(p)$ will recreate the interference frequency ζ_1, which is the low-frequency error of the SINS sensor. Under the assumption of the mathematical model of SINS error and observing the current error of the compensation scheme at the output of the filter $F(p)$, it is possible using parametric identification to correct the mathematical model errors in such a way that it recreates a time-varying current error component of the SINS. In summary, it can be stated that the mathematical models can use SINS error models, on which the Kalman filtering algorithms are based. Appropriate mathematical equations have been provided.

Extended Kalman Filtering Technique

Studies performed by the author on signal processing techniques indicate that the NN and EKF techniques are best suited for sensor fault detection and accommodation (SFDA) schemes, when different input signals from various sensors and various fault types are involved. Both techniques have been tested on a nonlinear UAV aircraft model. It is interesting to point out that the EKF technique is more effective in dealing with nonlinear model-based SFDA schemes and when such schemes strictly depend on mathematical description of the real-time operating system. The studies further indicate that the NN technique is particularly preferred, when adaptive structure and online training requirements are involved. In order to test the robustness to unknown inputs, different operating systems and noise measurements must be considered, especially in the case of a UAV model. Parameters with fluctuation values must be included in the EKF equations, if performance evaluation of such models and modeling errors are the principal requirements.

Types of faults include step-type, constant bias level, and hard and soft additive faults. It is justified to assume that a particular sensor can fail only once, which may or may not be proven later on. It can be further assumed that the faults are present in the pitch gyro sensor, when considering multiple sensor fault scenarios. Note that in order to reduce the false alarm rates and the number of undetected faults, a residual generation, padding and evaluation (RGPE) technique must be used.

Critical Steps for SFDA Test Conditions and Procedures: It is important to summarize critical steps for SFDA test conditions and procedures when dealing with a nonlinear UAV model. The following steps must be implemented:

- The EKF technique must be evaluated with emphasis on various nonlinear time-based equations involved.
- SFDA test conditions and procedures need to be identified when dealing with a nonlinear UAV aircraft model.
- For the sake of convenience, SFDA of a single-sensor fault should be considered.
- RGPE and residual generation and evaluation (RGE) approaches must be considered for residual processing and when NN technology is used for system modeling.
- In an NN-based SFDA scheme, NN training is switched off once a fault is detected.
- To maintain high system reliability performance, a sensor can fail only once and that failure is considered a permanent failure.
- Remember that the SFDA scheme is terminated once a sensor fault is detected and accommodated.
- In an NN-based RGE technique, NN is used for system modeling, while the RGE approach is used for residual processing.

Brief Physical Configuration of UAV Used in the SFDA Scheme: It is necessary to identify the UAV physical configuration vehicle based on the Eclipse class vehicle, which is typically powered by a small gas turbine power plant. In addition, the UAV configuration should identify three trailing edges, namely, aileron surface structure for roll control, flap-type device for altitude control, and elevator-type device for pitch control. Note that these devices are located on either side of the UAV center line. A rudder-type device, which is also located at the trailing edge of the fin, is not considered here. This particular device is used for yaw control.

Brief Description of UAV Model: A nonlinear decoupled six-degree-of-freedom model (which deploys open loop with no stability augmentation of UAV) should be considered for implementation in the Simulink environment. The UAV aircraft motion can be defined by a body axis system whose center of gravity is parallel to a fixed (inertial) earth axis. The UAV is assumed to be symmetric and rigid. Its flight dynamics are described by the standard 12 first-order differential equations, which include force, moment, kinematics, and navigation equations. The 12 principal variables include 3 components of velocity, 3 components of angular rates, 3 components of altitude, and 3 components of aircraft position relative to the earth axes.

Longitudinal Equations of Motion: In order to simplify the SFDA tests, only the UAV longitudinal motion should be considered, which includes roll, yaw, and sideslip angles each equal to zero. The remaining lateral motion variables should be assumed to be zero. Under these assumptions, the UAV aircraft motion is completely described by the axial force (X), normal force (Z), and pitch moment (M) as illustrated in Figure 4.8. It should be assumed that the forces and moments are strictly due to the aerodynamic, power, and gravitational effects.

These effects are functions of air density (p), UAV air speed (V_t), angle of attack (α), gas engine throttle setting (τ), and Z-axis coordinate of the engine thrust line (z_t), which is parallel to the ox_b relative to the aircraft body axis.

Expressions for the lift, drag, and pitching moment coefficients can be written as follows:

$$\text{The lift coefficient, } C_L = C_L\,(\alpha, \eta) \tag{4.42}$$

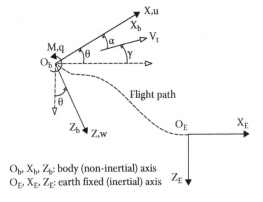

Figure 4.8 UAV orientation and relevant motion variables (shown in the *x–z* plane). Longitudinal equations of motion parameters: X, Axial force; Z, Normal force; Vt, Air speed; θ, Roll angle; ψ, Yaw angle; β, Side shifting angle; b, Body-axis parameters; E, Earth fixed axis parameters. (Graph from Samy, I. and Gu, D.W., *Fault Detection and Flight Data Measurement*, Springer-Verlag, Berlin, Germany, 2012.)

$$\text{The drag coefficient, } C_D = C_D\,(\alpha, \eta) \qquad (4.43)$$

$$\text{Pitching moment coefficient, } C_M = C_M\,(\alpha, \eta, q, dw/dt) \qquad (4.44)$$

where

α is the angle of attack

η is the trim angle

q is the angular velocity

w is the rotational speed

dw/dt is the rotational acceleration

Jet Engine Output Thrust: The engine output thrust is assumed to act only along the axial body axis of the UAV aircraft and can be expressed by the following equation:

$$\text{Output thrust available, } T_X = [\tau\, T(V_t, h)] \qquad (4.45)$$

where

τ is the throttle setting period

$T(V_t, h)$ is the available thrust, which is described by the look-up table

V_t is the air speed

h is the altitude

Longitudinal Trim: Initially all longitudinal UAV flight tests assume steady-state trimmed and rectilinear flight. Under these assumptions,

Table 4.2 Summary of UAV Model Trim Conditions

ANGLE OF ATTACK, α (°)	YAW ANGLE, γ (°)	AIR SPEED, V_t (M/S)	ALTITUDE, h (M)	THROTTLE SETTING, τ (CONSTANT)	TRIM, η (°)
7.55	0	32	1000	0.20	−12.98

Source: Samy, I. and Gu, D.W., *Fault Detection and Flight Data Measurement*, Springer-Verlag, Berlin, Germany, 2012.

all initial translational and rotational accelerations are zero, the initial sum of the forces and moments acting on the UAV aircraft are zero, and the initial angular velocities are also zero or parameter q is equal to zero. The UAV model trim conditions are summarized in Table 4.2.

The UAV flight varies at trim conditions, for all flight tests are defined and are summarized in the earlier table.

Unknown Inputs: In order to test the robustness of the SFDA scheme to unknown inputs, one must be able to incorporate such inputs in the UAV model. Three types of unknown inputs can be considered:

- Wind gust disturbances known as system noise. Wind disturbances can be modeled as zero-mean, white Gaussian gust disturbances acting on the angle of attack (α_{gust}) with a standard deviation of 0.10° and angle of sideslip (β_{gust}) with a standard deviation of 0.10°.
- The disturbance model is not considered an accurate model.
- Note that wind disturbances are always experienced during any aircraft flight.
- It is interesting to point out that α_{gust} is always in the direction of angle of attack orientation, which requires consideration of x and y components of the α_{gust}.
- Note that gyro noise should be neglected, if simplification of the analysis of the EKF-based or NN-based SFDA schemes is the principal requirement.
- If large amount of pitch gyro noise is included in the analysis, then the SFDA scheme will be susceptible to false alarms regardless of the accuracy of EKF or NN pitch rate estimates [7].

- Note that measurement of input known is essentially sensor noise, whereas parameter value fluctuation is recognized as signal with uncertainty.
- Parameter uncertainties should be considered in EKF equations but not in an NN-based SFDA scheme.

Summary

This chapter is dedicated to UAV navigation sensors and flight control system requirements. Innovation concepts for navigation sensors, aircraft control, error-free algorithms, and advanced propulsion systems are described in great detail for safe navigation and most reliable flight control. Existing sensors can be evaluated and redesigned for improved performance, optimum safety and reliability. Critical issues such as sensor performance levels, accurate navigation systems for UAVs, error-free algorithms, and advanced software development are given serious consideration with emphasis on UAV reliability and safety under combat environments. In the case of UCAVs operated by armed forces, additional sensors incorporating MEMS technology and nanotechnology may be required to meet stringent design requirements for weight, size, and power consumption. MEMS-based IMU requiring three-axis accelerometers, nanotechnology-based gyros, efficient and compact magnetometers, accurate barometric altimeters, and compact GNSS receivers are briefly described with particular emphasis on safety, reliability, and compact packaging. Advanced algorithm structures are reviewed for optimum reliability and safety. Note the algorithm uses a step-by-step procedure for solving the complex mathematical problems.

SINS, GNSS, and other navigation sensors are briefly described for improved navigation performance with emphasis on safety, accuracy, and reliability. Note SINS functions require four critical parameters, namely, correction of navigation route data, altitude data updates as needed, force transformation requirements for navigation path corrections, and velocity and position calculations for navigation path corrections. Quadratic spline approximations for calculations of incremental values are summarized involving vectors and matrices, incremental angles, time intervals, and UAV angular velocities. The mathematical procedure specified can lead to matrix and vector formats to represent the UAV altitude. Note the matrices used for altitude updates use the Euler angles. Specific forces are transferred to the navigation coordinate system using the altitude updates and are used for the correction of gravitational and centripetal accelerations. Equations for velocity and position calculations are provided for the benefits of graduate engineers and readers in general.

Algorithms for SINS are briefly mentioned. The SINS signals are provided by the GNSS system, magnetometers, and barometric altitudes using the EKF equations. Note that the vector quantities of SINS errors represent the UAV orientation, velocity, and position errors.

AFCS requirements for UAVs are summarized with major emphasis on vehicle reliability and navigation accuracy, while the UAV operates under AFCS. Critical functions of AFCS are identified with particular emphasis on reliability, safety, and navigation accuracy under all flight operating conditions and under combat environments. Note that robust design can significantly improve the AFCS needed for costly and complex UAVs, which might have uncertainties in its mathematical modeling and external disturbances due to aerodynamic conditions. Principal functions of the AFCS include handling of vehicle control, automatic stabilization of UAV angular positions along its three axes, automatic landing and takeoff, and automatic en route flight during the programmed conditions, which are described in detail.

Important functions of roll and yaw channels are specified with emphasis on the right amount of yaw damping and elimination of the slip when performing coordinated turns in automatic flight mode. Executing roles of longitudinal and lateral channels are identified.

Software requirements for AFCS are summarized with particular emphasis on robustness, reliability, and foolproof security for the UAV aircraft. Properties of specialized software for AFCS are identified with emphasis on optimum system performance, error-free performance, and adaptability, capability and ability of the software for interaction with AFCS elements as required. Overall performance-based specification requirements are briefly summarized.

Techniques to detect and isolate faults in UAV sensors are discussed with particular emphasis on unknown inputs, residual faults, fault isolation, fault accommodation, and fault analysis using computer modeling and feedback control analysis. The model-based FDI approach [8] is summarized, separating the faults in the UAV from manufacturing facilities. Some model-based FDI concepts are briefly described including the observer-based and Eigen structure–based methods. Benefits of EKF- and NN-based techniques are summarized. Solutions for INS system errors are identified. Importance of compensation filter is highlighted. Expressions for position errors and velocity errors are derived for the benefit of graduate students and readers in general. Major benefits of Simulink visual simulation software, a critical component of the MATLAB computer program, are summarized. Advantages of phase compensation of the UAV in SFDA are identified with emphasis on test conditions and test procedures. Important parameters of the UAV model are identified with particular emphasis on longitudinal equations of motion. Expressions for lift, drag, and pitching moment coefficients are derived in terms of angle of attack, trim angle limit, angular velocity, rotational speed, and rotational acceleration. Equation for jet engine output thrust is provided as a function of throttle setting, air speed, and UAV altitude. UAV model trim conditions are specified. Unknown inputs such as gyro noise, signal fluctuations, sensor noise, and wind gust are identified for vigorous modeling purposes. These unknown inputs can significantly affect the true UAV altitude, aerodynamic characteristics, and vehicle stability during the programmed flight.

References

1. V. Kharchenko et al., Urgent problems of UAV navigation system development and practical implementation, *IEEE Second International Conference Proceedings*, 2013, pp. 157–160.
2. A.A. Tunik, et al., Substantiation of requirements to UAV automatic flight control system development, *Actual Problems of Unmanned Air Vehicles Developments Proceedings (APUAVD), 2013 IEEE 2nd International Conference*, Kiev, Ukraine, 15–17 Oct. 2013, pp. 68–71.
3. M.L. Steinberg, Comparison of intelligence, adaptive, and nonlinear flight control laws, *Journal of Guidance, Control, and Dynamics*, 24, 693–699, 2001.
4. R.C. Nelson, *Flight Stability and Automatic Control*, 2nd edn., McGraw-Hill, New York, 1998, pp. 68–72.
5. V.B. Larvin and A.A. Tunik, On inertial navigation system error correction, *International Applied Mechanics*, 48(2), 213–223, 2012.

6. V. Kharchenko and S.I. Llnystka, Analysis of efficiency of algorithms of integrated inertial-satellite navigation system, *Mechanical Gyroscopic System*, 22, 32–43, 2010 (Ukraine).

7. N.K. Filyashkin and V.S. Yatskivsky, Prediction of inertial navigation system error dynamics in INS/GPS system, *Actual Problems of Unmanned Air Vehicles Developments Proceedings (APUAVD), 2013 IEEE 2nd International Conference*, Kiev, Ukraine, 15–17 Oct. 2013, pp. 206–209.

8. I. Samy and D.W. Gu, *Fault Detection and Flight Data Measurement*, Springer-Verlag, Berlin, Germany, 2012, pp. 5–8.

5

PROPULSION SYSTEMS AND ELECTRICAL SOURCES FOR DRONES AND UAVs

Introduction

This chapter is dedicated to power sources needed for drones, unmanned aerial vehicles (UAVs), and unmanned combat air vehicles (UCAVs). Drones may include miniature drones for small package or pizza delivery or delivery of commercial items as long as the weight and volume of the drones are within the specified limits. Power source requirements will be higher for attack drone and military missions providing intelligence gathering, reconnaissance, and surveillance (IRS) functions. Studies performed by the author reveal that nickel–cadmium and sealed zinc–silver are the most ideal power sources for small and microdrones with limited endurance and operating range capabilities. Drones with larger payloads and longer endurance may require fuel cells and propulsion sources to meet the power requirements [1].

Battery suitability must be given serious consideration, particularly for drones engaged in IRS functions and target tracking missions with limited payload capability. Fuel cells are available for heavy-duty, battlefield-bound UCAVs, which are equipped with side-looking radar, illuminating CW laser, and Hellfire missiles needed to destroy hostile targets.

Now, attempts are made to identify power sources for drones providing commercial services, micro-UAVs, tactical drones, and UAVs for IRS missions and fully equipped UAVs for combat operations. Sophisticated and complex designs have been developed for UCAVs known as hunter–killer UAVs. The power sources recommended or suggested for each of these drones or UAVs may or may not be adequate for the completion of the mission assigned, because the power requirements for vehicle or mission are strictly dependent on the vehicle payload, sensors involved, endurance capacity, and power consumption by the sensors aboard the UAV or drone [2].

Power Sources for Commercial Drones, Tactical Drones, and Minidrones

The power requirements for commercial drones and minidrones are very moderate. However, for tactical drones the power requirements will be slightly higher depending on the sensor's requirement. Again, the commercial drone power requirements are strictly dependent on the weight of the package or product to be delivered to the specified destination, the weight of the sensors involved, the en route number of stoppages, and the endurance capacity.

Electrical Power Sources for Commercial and Minidrones [3]

Studies performed by the author seem to indicate that the reliability, portability, and cost of electrical power sources for commercial drones and minidrones are of paramount importance. A compact gasoline-based internal combustion engine (ICE) can be used for battery charging, if needed for longer endurance capacity. The four-rotor copter shown in Figure 5.1 is a tactical drone and contains solid-state sealed nickel–cadmium batteries, which are found to be most economical. The hairline-like filament shown in Figure 5.1 carries a

Figure 5.1 Tactical drone capable of flying over long distances using a tether to provide communication link and power to the tactical drone.

low-loss fiber-optic cable that tethers the vehicle to a microcomputer and a battery pack. Note such a cable arrangement cannot be jammed or monitored by hostile operators. Sealed nickel–cadmium batteries are best suited for such a quadcopter, which is designed and developed by the DJI Company. Note that the batteries can be charged by the gasoline-based ICE, if required. These batteries can power a 1 kW CW laser and IR and GPS receivers. This battery pack weighs only 5–8 lb and can be used as a backup power source. This battery pack offers the best discharge and charge rates and acts as a reliable power storage device. Zinc–silver batteries are also available for such applications, where weight and size are of critical importance. This type of battery offers a specific power density of 1.8 kW/h. The critical components of a tactical drone include miniaturized IR camera, battery, and the 32 GB memory card.

As far as the reliability and unique capability of such a tactical drone are concerned, this type of performance is available to meet such performance date. Note that this type of tactical drone can look inside a building or over ridges before venturing elsewhere. It is capable of sending high-quality video signals and communication data and can set up a wireless communication relay for fighting soldiers. The 140 cm long quadcopter can hover at altitudes as high as 150 m (500 ft). Tactical drones would be most suitable for remote combat locations in hilly regions, which are difficult to reach when conducting military missions.

The quadcopter company has developed an 18 cm wide, 80 g hexacopter that can be easily slipped into a pocket. The tethered approach offers a unique option for the elimination of batteries [3]. Note that the company involved in the design of this particular tactical drone is engaged in the development of a hybrid propulsion drone, which carries a gasoline-powered ICE for a system where batteries are deployed.

Electrical Power Sources for Nano- and Micro-UAVs

The nano- or micro-UAV is the smallest version of the UAV. Essentially, their weight and physical dimensions are significantly lower compared to conventional UAVs. Dimensional parameters and technical specifications of renounced micro-UAVs are briefly summarized in Table 5.1.

Table 5.1 Dimensional Parameters and Technical Specifications of Micro-UAVs

MICRO-UAV DESIGNATION	WASP III	RAVEN RQ-11B
Wing span (ft)	2.375	4.5
Length (ft)	1.250	3.0
Vehicle weight (lb)	1.0	4.2
Micro-UAV speed (mph)	25–40	20–50
Operating altitude (ft)	50–1000	100–500
Operating range (mi)	3	6
Endurance capability (min)	45	60 90
Electrical power source	Battery	Rechargeable battery
Missions	Reconnaissance and surveillance	Target detection, convey security, battle damage assessment
Propulsion type	Two-blade propeller	Two-blade propeller

Note that the Wasp III micro-UAV flies reconnaissance and surveillance missions for smaller army units and its flight is controlled by the ground-station controller. Because of longer physical dimensions and use of a variety of sensors such as color cameras for achieving forward- and side-looking IR images. The Raven RQ-11B micro-UAV offers more mission functions compared to those from the Wasp III micro-UAV. Specific structural details of a hand-launched mini-UAV are shown in Figure 5.2. A California company has designed and developed high-reliability micro- and small drones using exotic manufacturing techniques. This company has developed Wasp III,

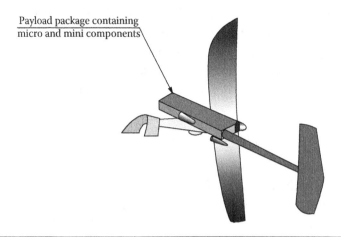

Payload package containing micro and mini components

Figure 5.2 Mini-UAV configuration can be launched by hand because total weight less than 4 lb.

Raven-11B, and PUMA AE nano- and microdrones with unique capabilities. In such vehicles, propulsion power can be supplied by gasoline-based or kerosene-based ICEs, while the electrical power is provided by the battery packs. Battery suitability is the most serious issue.

Battery Suitability

Practically, most of the small UAVs normally use battery packs with minimum weight and size. Electrical engineers should access a number of battery parameters when evaluating the suitability of the battery type for a particular application. The most common factors used in battery analysis include nominal voltage, energy capacity, energy density, self-discharge capability, and dynamic considerations. The nominal voltage is the voltage as measured across the battery terminals. Engineers should use multiple batteries in series and parallel configurations to achieve the required voltage or current for the electrical power supply. Note that energy capacity is the stored energy content of the battery and is expressed in joules.

Reliability Aspect of Batteries Is Vital for Electronic Drones: Reliability and ability of the battery to provide constant voltage regardless of electrical load are of critical importance and strictly depend on battery chemistry. However, the battery chemistry relies on electrochemical reactions to provide electrical energy. Some of these reactions are more potent than others, which can lead to the development of small batteries in a very efficient way with minimum cost and complexity. The size-to-energy ratio is the electrical density capability of the battery. As a matter of fact, the higher the energy density, the more costly will be the battery technology. Because of this, battery design engineers constantly struggle to find the optimum balance of cost versus energy density.

A battery does not last forever. Even as batteries sit on the self, electrochemical reactions still take place, which slowly diminish their energy content. This naturally occurring process is known as self-discharge rate. This process affects the service of alkaline batteries, which ranges from 7 to 10 years. Standard lithium batteries have a service life of 10 to 15 years. But lithium thionyl chloride ($LiSOCl_2$) batteries have life in excess of 40 years. Note that self-discharge rates and other deteriorative mechanisms seriously affect battery life. Furthermore, excessive temperature and duty cycle seriously affect the deteriorative mechanisms. In addition, fluctuating duty cycle requirements have adverse effects on the discharge characteristics of the battery and ultimately on its reliability.

Dynamic physical parameters such as variations in operating temperature, output impedance, duty cycle, and energy delivery technique can affect load conditions and ultimately shape the battery selection process. Studies performed by the author seem to indicate that the battery discharge profiles require special attention from the battery design engineers. Operational considerations such as continuous operating temperature can seriously affect the battery performance and its reliability. System-level consideration such as battery interval replacement and system-level voltage requirement can affect the cost and complexity.

Battery Replacement: This is a costly issue. Sometimes, battery replacement becomes a critical reliability issue, particularly for field operating equipment or an onboard UAV-based system. In such situations, zinc–air batteries have been found suitable because of their high energy density close to 1.69 MJ/kg and ability to deliver high peak currents for a short duration not exceeding 3 months or so. It is interesting to mention that nickel–cadmium batteries deployed in the 1990s era UCAV Raven could keep it aloft for only 30 min. Note that for longer airborne operations, high-performance batteries like $LiSOCl_2$ will be found most ideal. In essence, batteries such as $LiSOCl_2$ are best suited for high-performance, armed heavy-duty UAVs as shown in Figure 5.3, where reliability, longevity, and consistent performance under extreme environments are the principal requirements to meet IRS missions. High-performance UCAV platforms like Predator B shown in Figure 5.4 heavily armed with Hellfire missiles and offensive weapons and fully equipped with electro-optical (EO) and electromagnetic (EM) sensors would require a propulsion system capable of providing enough thrust under all flight conditions while engaged in classified missions.

Performance Capabilities of $LiSOCl_2$ Batteries: Such batteries have demonstrated high reliability, high service life exceeding 40 years, and consistent performance under extreme temperature environments.

Figure 5.3   Armed UCAV aircraft capable of undertaking intelligence gathering, reconnaissance, and surveillance missions.

Figure 5.4   High-altitude, high-performance armed UAV configuration best suited for undertaking classified missions.

Lithium is the highest nongaseous metal that offers the highest specific energy or electrical energy per unit weight and maximum energy density or energy per unit volume among all available battery chemistries. Furthermore, lithium cells use a nonaqueous electrolyte that is highly desirable for long service life and optimum reliability under harsh thermal and mechanical environments. Because of these qualities, $LiSOCl_2$ batteries are best suited for heavy-duty UAVs conducting complex and dangerous military missions over long durations in hostile territories. Their unique characteristics can be summarized as follows:

- Open-circuit voltage (OCV) between 2.7 and 3.6 V.
- *Extreme temperature limits*: -55°C to +125°C.
- *Battery performance at cryogenic temperature*: 80°C.
- Batteries tested down to cryogenic temperature at -100°C and remained operational.
- *Operation life*: greater than 40 years.
- These batteries are constructed using high-quality materials and advanced manufacturing techniques, which significantly reduce the electrolyte leakage or short circuits.
- These batteries have demonstrated low annual self-leakage at ambient temperatures.
- Under dormant-mode operations, these batteries have demonstrated improvement in service life.
- It is easy to implement power management solution based on specific applications, particularly when high reliability is of major concern.
- It is easy to couple energy-harvesting devices with the rechargeable lithium thionyl chloride batteries, which store the harvesting energy.
- Consumer-grade $LiSOCl_2$ batteries are not suitable for remote wireless applications.
- These batteries can be charged directly from a dc source or by energy-harvesting devices.
- These batteries can be used to power the sensors aboard the UAV platforms, if the battery capability is adequate to meet the sensors' electrical power requirements.
- This particular battery uses $SOCl_2$ as a liquid cathode, which offers several advantages such as high energy density capacity close to 1410 Wh/kg, discharge rates as high as 200 mA/cm^2, and discharge durations exceeding 25 min.
- For a heavy-duty UCAV such as Predator B, such batteries may not be suitable to meet excessive power requirements for

Table 5.2   Power Output Levels as a Function of Discharge Duration and Using Optimized Electrolyte

DISCHARGE DURATION (MIN)	OUTPUT OF 20 kW	OUTPUT OF 10 kW	10 kW BATTERY WITH OPTIMIZED ELECTROLYTE
2	13.5	6.8	7.8
4	14.1	7.3	8.1
6	14.1	7.5	8.2
8	14.1	7.8	8.3
10	13.9	8.0	11.6

sophisticated EO and EM sensors. In such cases, fuel cells can be deployed [2].

• Power output levels of such batteries as a function of discharge duration and using optimized electrolyte are summarized in Table 5.2.

Compact or Miniaturized High-Capacity Batteries for Commercial Drones   This section will focus on miniaturized batteries best suited for commercial drones and micro-UAVs for short endurance missions. Commercial drones or micro-UAVs with short-duration missions could perform missions with moderate- or high-capacity miniaturized batteries. This approach will significantly reduce the weight and size of the power sources, and batteries will be most ideal for such airborne vehicles. Also, such vehicles normally deploy tiny gyroscopes, accelerometers, simple air speed sensors, and compact GPS receivers, and therefore, the electrical power requirements will most likely be moderate for small UAVs. Even U.S. military planners are able to work out detailed tactics, techniques, and procedures for battlefield soldiers to use small UAV or micro-UAVs in combat environments. This is possible due to the availability of zinc–silver (Zn–Ag) and nickel–cadmium (Ni–Cd) batteries and battery packs. Note that Zn–Ag batteries suffer from high procurement costs and poor power performance at low temperatures but such batteries have high longevity. Note most of the micro-UAVs or small commercial drones do not operate at altitudes greater than 500 ft, and hence, poor performance at low temperatures is out of question. Atmospheric temperature (T) in the absolute thermodynamic scale can be calculated using the following equation:

$$T = T_M \left(\frac{M}{M_o} \right) \tag{5.1}$$

where

T is the atmospheric temperature (K)

T_M is the molecule temperature (K)

M is the molecule weight scale (no unit)

M_o is the sea level value of the molecule's weight

Calculations at sea level

Parametric values must be computed at specific height or altitude using sea level values, which are as follows:

T_M = 288.16 K at sea level (H = 0 m)

T = 288.16 K with atmospheric temperature of (273–288.16) = 56.34°C

Alternately, one can look for atmospheric temperature values at specific height or altitude.

Sealed Ni–Cd Battery for Micro-UAVs and Small Drones: Comprehensive research studies undertaken on such batteries seem to indicate that Ni–Cd is the most ideal for micro-UAVs and electronic drones intended for the delivery of commercial products at a short distance, not to exceed 5–10 mi. The operating life of Ni–Cd batteries ranges from 10 to 15 years with no compromise in optimum reliability and electrical performance. A battery pack comprising such sealed batteries weighs between 5 and 8 lb and is capable of providing 1 kW of electrical power sufficient to power the IR laser, IR receiver, compact GPS receiver, and miniaturized navigation system. This type of battery pack has been deployed by micro-UAVs and electronic drones to meet their mission requirements and endurance capabilities. Such sealed batteries can be used as backup power sources, which offer the best discharge and charge rates and a reliable power storage source. The research studies performed by the author also indicate that zinc–silver (Zn–Ag) batteries are also best suited for applications where minimum weight, size, and optimum reliability are the principal design requirements. Such battery pack offers a specific power density of 1.8 kW/h. However, such batteries suffer from high procurement costs and poor performance particularly at low temperatures. But Zn–Ag batteries can power the sensors aboard micro-UAVs, provided the vehicles do not fly more than 2500 ft where the vehicles will experience moderate atmospheric temperatures. In summary, it can be stated that the Zn–Ag battery pack offers an output voltage from 1.5 to 1.8 V, excellent reliability over operating temperatures from 20°C to +60°C, life cycle of 50–100, specific energy capability from 90 to 100 Wh/kg, and self-discharge rates less than 5% per month.

Critical Role Played by Ni–Cd Batteries in UAVs   Drones or micro-UAVs using single Ni–Cd batteries can remain aloft in the air for a limited period not exceeding 30 min. However, the vehicles can remain aloft in air for longer periods if a battery pack consisting of 5–6 batteries is deployed.

The Dragon Eye UAV designed and developed by Naval Electronics Laboratory (NEL) has demonstrated remarkable performance in Iraq. This vehicle weighs only 5 lb and can be assembled and stored in a small suitcase. This microvehicle is used as a stealth information gathering source. Note that the Dragon Eye is equipped with an electric motor operated by a single Ni–Cd battery. This vehicle can fly nonstop for a period of 50 min and can operate over a range exceeding 40 mi while flying at an altitude close to 10,000 ft.

Even large-size UAVs such as RQ-1 deploy Ni–Cd rechargeable battery packs as a standby power source. Each of the rechargeable battery pack weighs 8 lb. RQ-1 also uses a dedicated power generating system similar to a conventional commercial automotive alternator. This type of backup power sources offers an iron-clad reliability while engaged in critical mission tasks.

Fuel Cells for Heavy-Duty UAVs   Research studies performed by the author [2] reveal that fuel cells are most suitable for applications where an uninterrupted supply of electrical power over a specified duration is essential with no compromise in reliability, performance, and continuous operation of sensors aboard the UCAVs. In other words, fuel cells are of critical importance where uninterrupted operation, secured data collection, and completion of critical military missions are crucial. Fuel cells are most attractive in terms of cost, weight, and size as long as the power consumption is not excessive. In summary, fuel cells are mostly attractive for heavy-duty electronic drones such as UAVs, which provide intelligence gathering, reconnaissance, surveillance, and target tracking capabilities over battlefields for the benefit of field commanders. Fuel cells are capable of providing power simultaneously to side-looking radars, EO sensors, microwave and infrared sensors, communication receivers, and laser illuminators for missile guidance. Performance capabilities and major advantages of fuel cells can be summarized as follows:

- Fuel cells are capable of providing continuous electrical power over a period of time ranging from 5 to 10 h, which is needed by the UCAVs to collect the needed data for making military decisions.
- Fuel cells are considered most attractive in portable electrical sources, where minimum weight, size, reliability, and availability of uninterrupted electrical power are the principal operational requirements.
- In addition to the use of fuel cells for UAVs or UCAVs, there are several applications where fuel cells can play important roles including logistic and cargo transportation, security inspection at the borders, removal of wounded soldiers, supply of essential items in remote areas, and destruction of mines and improvised electronic devices.
- Fuel cells are best suited for UAVs deeply involved in undertaking reconnaissance, surveillance, and target acquisition (RSTA) missions.
- Successful and precision target location and missile attack by the pilotless electronic drones or UCAVs powered by fuel cells will be the preferred way to fight future military conflicts and terrorist activities and undertaking counterintelligence missions in hostile and conflict regions.
- To undertake critical missions with duration ranging from 4 to 8 h, deploying a number of EO, microwave, and infrared sensors and carrying higher payload would require an exact power output capability of the fuel cell for a UAV or UCAV platform.
- Research studies undertaken by the author seem to indicate that mission duration, task assigned, and payload capacity will ultimately define the power output capability of the fuel cell.

- These studies further indicate that the closed-cycle design of a fuel cell is best suited for deployment in UAVs or UCAVs.
- Fuel cells have been deployed in space applications as early as the 1960s and have demonstrated remarkable performance and reliability over extended periods.
- Fuel cells have been recognized as most ideal for buses, scooters, hybrid electric cars, and automobiles where constant electrical power, reliability, and safety are the basic requirements.
- High-capacity ion-exchange membrane (IEM) fuel cells were developed by NASA scientists in 1960 for the Gemini program to demonstrate the feasibility of fuel cell deployment in space applications.
- A power module consisting of fuel cells and using methanol as fuel to generate hydrogen and liquid oxygen as an oxidant has demonstrated fuel cell application in submarine propulsion systems. This particular fuel cell module demonstrated a volumetric power density in excess of 2.5 kW/ft^3, quick start-up capability at 90% full power availability, and power source efficiency better than 65% under severe space environment with optimum reliability.
- Hydrogen–oxygen fuel cells using porous silicon structure and proton-exchange membrane design shown in Figure 5.5 will generate electrical power with optimum reliability and safety and minimum cost and complexity.

Important thermodynamic efficiency and OCV (E^{00}) for various fuel cell types are summarized in Table 5.3.

A comprehensive examination of the aforementioned data is that a fuel cell using zinc offers a higher open-circuit cell voltage close to 1.56 and highest thermodynamic cell efficiency of 96.8%. With the conversion of chemical energy into electrical energy, additional energy is gained due to absorption of heat energy from the surroundings. In addition, when the reaction entropy is negative, the fuel cell efficiency is less than 100% irrespective of the fuel used. That is why hydrogen offers the lowest thermodynamic efficiency or the ideal cell efficiency as indicated in the first row on Table 5.3. Because of high volumetric power density, fuel cells are best suited for UAVs and

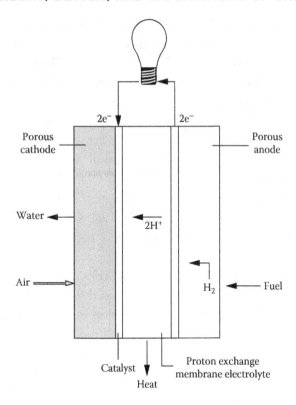

Figure 5.5 Critical elements of a high capacity fuel cell using PEM design architecture, porous anode and cathode electrodes.

Table 5.3 Thermodynamic Efficiency and Open-Circuit Cell Voltage for Various Fuel Cells

CELL TYPE USING	VALENCE	E^{00} (V)	THERMODYNAMIC CELL EFFICIENCY (%)
Hydrogen	2	1.229	83.2
Carbon oxide	2	1.066	90.8
Formic acid	2	1.480	91.5
Methanol	6	1.214	96.7
Methane	8	1.060	91.8
Ammonia	3	1.172	88.5
Hydrazine	4	1.558	96.8
Zinc	2	1.657	91.3

Source: Jha, A.R., *Next Generation of Batteries and Fuel Cells for Commercial, Military and Space Applications*, CRC Press, Boca Raton, FL, 2012, pp. 112–113, 259–261.

UCAVs. In essence, fuel cells will occupy the least space in UAVs, which is of critical importance.

Fuel cells' power requirements strictly depend on the mission duration, payload capacity, electrical power needed for the sensors aboard the vehicle, and tasks to be performed. Preliminary studies performed by the author for the fuel cell power rating seem to indicate that fuel cells with power ratings between 3 and 5 kW may be adequate for missions not exceeding 3–5 h. Additional power close to 1 kW may be required if side-looking radar and exotic sensors are deployed by the UAVs involved in combat operations, where precision target tracking and deployment of multiple sensors are the principal tactical mission requirements.

Studies undertaken by the author seem to indicate that lithium thionyl chloride ($LiSOCl_2$) chemistry-based batteries can operate over 40 years, and energy-harvesting devices coupled with special rechargeable lithium-ion batteries designed for extreme temperature environments can also deliver up to more than 20 years of battery life. $LiSOCl_2$ chemistry is best suited for use in extreme temperature environments. It is equally important to mention that lithium is the lightest nongaseous metal that is capable of offering high specific energy or the highest energy per unit weight and highest energy density or highest energy per unit volume among all available battery chemistries.

In addition, all lithium cells use a nonaqueous electrolyte and have normal OCV between 2.7 and 3.6 V. Note that the absence of water allows certain lithium batteries to operate under extreme temperature environments ranging from –55°C to 125°C. Some lithium battery designers have placed the batteries in the cryogenic chamber, subjected them to progressively lower cryogenic temperatures down to –100°C, and the batteries have retained all performance capabilities. Some lithium manufacturers are claiming that superior-grade $LiSOCl_2$ batteries have been designed and developed using high-quality materials and advanced manufacturing techniques. Such techniques have demonstrated reduction in electrolyte leakage and avoidance of short-circuit operations. In other words, deployment of high-quality materials and advanced manufacturing techniques has demonstrated increase in battery life and optimum reliability under extreme thermal and harsh operating conditions.

LiSOCl$_2$ Batteries for Next-Generation Tactical Drones and Micro-UAVs Operated by Military Authorities: Near silent operation in the air and on the ground and absence of gaseous emissions are of great tactical importance particularly in military applications. If a gasoline-based internal consumption engine or gas turbine is used by the UAV or electronic drone, its stealthiness is compromised based on the noise generated by the propulsion system. To avoid such detection of the UAVs or drones, alternative electrical power sources such as high-capacity LiSOCl$_2$ batteries will play a critical role in the protection and security of tactical drones. Note that these batteries have high electrical energy capabilities that can be used over extended periods, which is not possible with other types of electrical power sources. Furthermore, these rechargeable batteries are suited for tactical drones and micro-UAVs because they meet minimum weight and size performance parameters.

Deployment of Ni–Cd Sealed and Rechargeable Batteries for Micro-UAVs, Commercial Transports, and Military Aircrafts: Research studies undertaken by the author indicate that sealed and rechargeable Ni–Cd batteries have been deployed not only by tactical drones and micro-UAVs but also by commercial and military aircraft. Sealed Ni–Cd batteries have been developed by Acme Electronics Corporation and Eagle-Picher Industries of the United States. Rechargeable and sealed Ni–Cd batteries are currently deployed in commercial transports such as MD-80, MD-90, DC-9, and Boeing 777. In addition, some F-16 fighter aircraft and Apache helicopters are using these sealed and rechargeable batteries. Sealed and maintenance-free Ni–Cd batteries have been approved for military aircraft including F-16, F-18, B052, and E-8 advanced Airborne Warning and Control System (AWACS) aircraft. Market performance charts indicate that no other battery can meet the battery capability of a vented Ni–Cd rechargeable battery in starting the aircraft engine at −45°C ambient temperature without fail. Such is the impressive and reliable performance record of Ni–Cd rechargeable batteries. The outstanding performance capabilities of these batteries can be summarized as follows:

- *Specific energy*: 45–65 Wh/kg
- *Energy density*: 150–200 Wh/L
- *Cycle life to 80% capacity*: 1100–1500

- *Self-discharge at 20°C/month*: 10%–15%
- *Fast charging time*: 1 h
- *Nominal cell voltage*: 1.20 V
- *Operating temperature*: –20°C to + 60°C
- *Procurement cost*: 0.5 $/Wh
- *Overcharge tolerance*: Moderate
- *Power density*: Excellent
- Vented Ni–Cd batteries meet stringent system reliability under severe thermal and mechanical operating environments

Power Sources for Drones, Electronic Drones, and Micro-UAVs

This particular section will deal with various power sources needed to operate electronic drones, micro-UAVs, and full-size UAVs. Mechanical power driving sources are essential for drones and UAVs to operate from one point to another in the atmosphere. Drones and micro-UAVs can be operated by gasoline-based ICEs coupled with a two- or three-propeller mechanism. Commercial drones and quadcopters can be operated using suitable commercial batteries and micropropeller mechanisms, and the control of these can be managed by the operator on the ground. The type of power sources strictly depends on the drone or micro-UAV type, payload, sensors deployed aboard the vehicle, operating range, and endurance capability.

Propulsion Sources for Electronic Drones and Quadcopters

Commercial electronic drones and quadcopters (Figure 5.1) typically operate at lower altitudes not exceeding 500 ft. Rechargeable batteries are available to maneuver these vehicles using electronic commands from the ground control operator. A sophisticated tactical drone capable of flying over long distances uses a tether to provide communication signals and electrical power from the battery to the vehicle. Such vehicles can make frequent stops to deliver products to customers at their specified addresses. Gasoline-based ICEs will neither be suitable nor economically feasible to operate these copters for specific applications.

Mini-UAVs weighing less than 3 lb or so can be launched by hand as illustrated in Figure 5.2. In other words, no propulsion system is required to launch mini-UAVs in the lower atmosphere, if the vehicle load is less than 3 or 4 lb.

UAVs involved in undertaking intelligence collection, reconnaissance, surveillance, and target tracking missions (Figure 5.3) need a propeller-based propulsion system driven by a gasoline-based ICE or mini-jet engine to keep the vehicle aloft in the atmosphere comprising three-dimensional coordinate space. The number of propellers required is dependent on the thrust requirement, top speed of the vehicle, and atmospheric conditions. These propellers can be driven by batteries or gasoline-based ICE or micro-turbojet. The selection of the propulsion system should consider several critical issues such as cost, weight, and size of the system, speed of the vehicle, and ability to meet the peak endurance capability.

Gasoline-Based ICEs for Small Drones: Studies performed by the author on ICEs seem to point out that such engines coupled with propellers are best suited for small UAVs capable of undertaking short border reconnaissance missions over limited durations with minimum cost and complexity. ICE efficient two-stroke engine design, four-stroke engine design, or rotary engine design configurations are available in open markets.

Note a rotary engine does four distinct job functions, but each of them happens within the same housing section, while the piston moves continuously from one position to the next. The rotary engine was originally designed by Felix Wankel and thus known as the Wankel rotary engine. Rotary engines are best suited for undertaking reconnaissance and surveillance missions over hostile and unknown lands, because such engines produce minimum noise. Furthermore, its infrared signature of a two-stroke engine is much less than that from the four-stroke ICE because its exhaust contains relatively less hydrocarbon content. Note that ICEs work on the principle of injection theory, which includes the following four types of injection systems:

1. Gasoline injection system
2. Diesel injection system
3. Methane injection system
4. Magnesium injection system

Highlights of ICEs

- An ICE comes with reciprocate engine or rotary engine.
- Two-stroke gasoline-based ICE costs between $50 and $65 and has efficiency better than 50%.
- Four-stroke gasoline-based ICE costs between $60 and $90 and has efficiency close to 43%.
- In a gasoline-based ICE, a mixture of gasoline and air generates the power.
- A closed-cycle ICE is more efficient.
- ICEs for deployment in UAVs require well-designed bearings, gaskets using advanced materials, and valve seats.
- Setting of appropriate clearances is absolutely necessary to retain high engine performance.
- After the ICE assembly, compression ratios must be checked to ensure engine performance over long duration.
- The coolant system is pressured to 30 lb/in.^{2} for 30 min to ensure optimum engine performance.

Propulsion Systems Using Gas Turbines and Jet Engines for UAVs: It is critically important to mention the basic operational concepts of pure jet engines and gas turbine jet engines. In each case, the exhaust jet flow leaves at the rear end of the engine and creates the forward thrust, which moves the UAV or aircraft in the forward direction. Note that UAV movement is controlled by the aileron, vertical tail, and wingtip surfaces.

Essentially, a jet engine is a reaction engine discharging a fast-moving exhaust jet that generates the thrust by jet propulsion in accordance with Newton's laws of motion. The intake air flows through a jet engine combustion chamber and mixes with the fuel in the chamber to convert the jet fuel and air mixture to burn. This leads to generation of a tremendous thrust that moves the UAV aircraft in the forward direction.

Gas Turbine Jet Engines: Most commercial jet transports are powered with turbofan engines, which are also known as gas turbine engines. The air enters from the front of the engine and is compressed by a multistage compressor. The fuel and the compressed air mixture burn in the combustion chamber and the hot gases exit the chamber from the rear through an exhaust manifold. The gas turbine blades are generally made from MuMETAL®, which retains high tensile and compressive strengths at elevated temperatures.

The gas turbine uses an aviation fuel, which is high in octane with very high calorific contents. Studies performed by the author on the benefits of gas turbine engines seem to reveal that they are best suited for applications where high speed and output energy are the principal requirements.

Suitability and Deployment of Appropriate Sources for UAV Propulsion

In this section attempts will be made to justify the suitability and deployment of specific power propulsion sources for micro-UAVs, medium-size UAVs, and full-size UCAVs. It is of paramount importance to mention that weight, size, and operating life of the proposed propulsion source are the most stringent requirements for incorporation in a specific UAV platform.

Propulsion Systems for Micro-UAVs and Commercial Electronic Drones

As mentioned earlier, the type of propulsion system is strictly dependent on the UAV type, sensors aboard the vehicle, endurance of the mission, type and nature of mission, operating range and altitude, and weight and size restrictions. Since operating height, operating range, and endurance duration requirements are generally moderate for micro-UAVs and commercial electronic drones, a conventional propulsion system is not essential. For such platforms, a Ni–Cd or lithium battery pack with a gasoline engine–driven generator set will be most suitable to propel micro-UAVs and commercial electronic drones or quadcopters equipped with quadrotors as illustrated in Figure 5.1. A gas engine–driven generator configuration is generally not recommended because its extra weight and size may not be acceptable by the UAV. Note that these rotors will be electronically controlled by the ground operator to maintain the vehicle flight profile. The ground control operator will stop to deliver payloads at the desired locations and resume the vehicle flight using the electronic command signals from the ground control station for the next destination or assignment.

Some commercial electronic drone designers plan to use hybrid propulsion systems currently in development phase. Such a hybrid system will carry a powered microgenerator that will charge the batteries during the flight. The new copter will be equipped with six 26 in. rotors and will fly nonstop more than two more h in gusts of up to 35 mph while carrying payloads close to 5 lb. This particular copter will be equipped with gimbal-stabilized IR camera, extra battery, and 32 GB memory card and reader. Recent market reports have indicated that a large increase in the use of such UAVs is expected in the near future for three separate platform categories including personal/toy drones, small commercial drones, and commercial large drones using hybrid or pure electric vehicles. Note that hybrid drones will be most suited for delivering the products urgently needed for specific applications.

Future Market Forecast for Hybrid or Electronic Drones   Recent market research reports reveal that all markets will significantly grow for all categories of drones, whether hybrid or pure electronic drones, with net sale exceeding close to $5 billion by year 2025 in satisfying the following benefits, needs, and applications:

- Agriculture.
- For the delivery of urgently needed commercial products.
- Growing percentage of electronic drones needed for fast delivery of medicines to hospitals.
- Significant use of pure electronic drones is expected in the near future because of minimum cost and complexity.
- Near silent operation both in air and on the ground is most valuable for both military and civil applications.
- Free from gaseous emissions.
- Virtually maintenance-free operation.
- Stealth design with minimum radar cross section (RCS).
- Impossible detection by enemy radar because of small-vehicle RCS and low altitude operation.
- Rapid deployment of quadcopters is possible due to minimum cost.
- Latest market survey confirms that the growing percentage of the total market forecast will be electronic drones because of minimum cost and complexity.

- Cost of components deployed by small UAVs or drones has been dropping sharply due to large increases in volumes.
- Future improvement in design complexity and sophistication will significantly enhance the electronic drone performance capabilities.
- Electronic drone designers are predicting a bright future for both hybrid and pure electronic drones in undertaking both civil and military missions.
- Since the size of the drone industry is rapidly growing, the safety and security agencies are pushing for traffic control for these unmanned vehicles. Note that NASA has been researching traffic management needs for immediate federal legislations.
- Drones with hybrid propulsion technology will offer higher payload capacity, which means a large quantity of items can be delivered in one trip [4].
- Such drones are in advanced development phase and will be available in the near future.
- Lightweight gasoline-powered motor–generator sets soon will be available, which can charge the onboard batteries in flight.
- Studies performed by the author on flight capabilities of various drones reveal that quadcopters can fly only a few h a day while carrying deliverable payload less 2 lb.
- Amazon is planning to deploy such electronic drones to deliver small packages to their customers in minimum time.
- Amazon plans to fly their quadcopters at an altitude not exceeding 400 ft and speed less than 100 mph.
- Amazon officials state that their copters will be equipped with lightweight gimbal-stabilized IR cameras, extra battery pack, 32 GB memory card and reader, and compact GPS device. According to Amazon officials, using these onboard equipment and devices, packages will be delivered to their customers in minimum possible time.
- These quadcopters will be electronically controlled by well-trained ground control operators to minimize flight time to their customers.
- Electronic drone designers predict that the size of the nonelectronic drone market will be exclusive for military applications.

Table 5.4   Technical Specifications for Remote Control Rotary-Wing UAV

UAV COMPONENT DESCRIPTION	SPECIFICATION
Wing span (m)	1.6
Operating altitude (m)	2000
Maximum takeoff weight (kg)	15 (33 lb)
Maximum speed (km/h)	30 or 18 mph
Payload (kg)	5 or 11 lb
Endurance (min)	50
Electrical propulsion source	Lithium battery
Principal function	To provide real-time video
Radio control range (km)	3–5

- As a matter of fact, there will be a rapid transition from non-electric drone technology to UCAV deployment by the armed forces.

Electronic Drone Classifications: Electronic drones have been designed with four rotors, which are known as quadrotor drones. Electronic drones with six rotors are available to heavy payload and long endurance requirements. Technical articles presented by various authors on electronic drones seem to reveal that these drones are widely used to provide search, patrol, and surveillance missions as well as for agriculture and disaster monitoring operations. High-capacity Ni–Cd rechargeable battery packs have been deployed for electronic drones. Lightweight and compact gasoline-based ICEs can be used to charge these batteries during long-endurance missions. High-efficiency thermal insulation or nanosilica aerogel is used to provide thermal insulation and low acoustic noise specifications, particularly for remote control, six-rotary-wing UAV aircraft with the technical specifications shown in Table 5.4.

Propulsion Systems for Full-Size UAVs and UCAVs

Powerful propulsion systems will be needed to meet the size, weight, and payload requirements for UAVs and UCAVs. UAV propulsion system power requirements will be small, if it is intended to undertake intelligence, reconnaissance, and surveillance missions for short durations not exceeding 2–3 h. However, propulsion system power requirements will be moderate, if the aforementioned UAV is to undertake armed missions. Finally, the propulsion system power requirements will be very high for the UCAV, because this particular vehicle is required to carry a large number of sensors, a variety of offensive weapons, and defensive RF, IR, and EO sensors to meet the mission performance requirements.

Categories of Propulsion Systems

It is appropriate to describe various categories of propulsion systems and their performance capabilities for the benefit of the readers. Research studies performed on potential propulsion systems for deployment in UAVs and UCAVs indicate that the following types of engines and turbines are available to provide thrust levels needed for UAVs and UCAVs:

- Small jet engines
- Small gas turbines
- Small turbine engine
- Compact gas turbine engines
- Miniaturized jet engines
- Tesla turbine engines
- Turbofans that are available for UAV applications
- High-thrust turbofan jet engines

The aforementioned jet engines come in different sizes and with various thrust ratings.

Distinction between Combustion Turbines and Jet Engines [5]
Gas Turbine: A turbine is called a gas turbine because it compresses the gas, usually the air. There are three critical stages for a gas in the turbine. In the first stage, the air is compressed by the multistage compressor that raises the temperature of the compressed air. In the second stage, the fuel is added to the combustion chamber. In the third stage, hot air travels through the blades of the multistage turbine that produces the thrust when the hot air exits the tail pipe. Note that the turbine drives the compressor because it is rigidly coupled to the turbine.

Jet Engine: A jet engine takes a large volume of hot gas from a combustion chamber that is a part of the gas turbine. Note that the jet propulsion uses either solid or liquid propellants. Note that a ramjet is essentially produced when the gas turbine feeds the hot gases through a nozzle that accelerates the speed of the jet. Also note that as the jet accelerates through the nozzle, its speed is significantly increased, thereby increasing the thrust to a higher level. A jet is further classified into various categories depending on the speed of the jet stream:

- Subsonic jet when its speed is less than 1 M
- Supersonic jet when its speed is greater than 1 M
- Ramjet when its speed is less than 3 M (approx.)
- Scram jet when its speed is greater than 5 M (approx.)

In the case of a fighter jet when it is operating in afterburner mode, the jet aircraft is flying at a speed ranging from 2.5–3 M.

Jet Engine Classifications: Classifications of jet engines are identified in the following distinct categories:

- Turbojet engine
- Turbofan engine
- Ramjet engine
- Scramjet engine
- Rocket engine

Critical Performance Parameters of Jet Engine and Gas Turbines: The performance parameters for jet engines and gas turbines can be summarized as follows:

- Energy efficiency
- Fuel calorific value
- Fuel consumption per hour
- Power output-to-weight ratio
- Thrust-to-weight ratio
- Thrust lapse for jet engine
- Overall pressure ratio for jet engines and gas turbines

Research studies undertaken by the author on the suitability of propulsion systems for large UAVs and UCAVs indicate that turbojet and turbofan jet engines must meet the thrust requirements compatible with endurance durations under prescribed flight conditions. Note that miniaturized versions of these jet engines and gas turbines are available for moderate thrust requirements.

Propulsion Systems for UAVs: Selection of a propulsion system is strictly dependent on the weight and size of the vehicle, missions to be executed, mission durations, overall payload, UAV thrust requirement, sensors aboard the vehicle, endurance capability, operating range, and altitude.

If the armed UAV is required to undertake IRS missions, the propulsion system requirement could range from low to moderate depending on the number of EO and EM sensors, laser-guided Hellfire missiles, and other offensive weapons involved. To undertake the first three IRS missions, a small turbojet with low thrust capability may be found most suitable. Studies performed by the author on various propulsion sources seem to indicate that gasoline-powered engine motor–generator sets will not be able to meet the propulsion system requirements for such missions. It should be further noted that even the medium gas-powered jet propulsion system will not be appropriate to meet the propulsion requirements because of excessive weight and size. The most serious drawback using a gas turbine or a turbojet is its excessive weight, size, and high noise level generated by these sources. Such noisy propulsion systems are not appropriate for undertaking reconnaissance and surveillance missions or any military mission requiring absolute secrecy. In summary, UAVs designed for undertaking IRS missions must meet the stringent propulsion system performance requirements with particular emphasis on endurance duration, sustainable thrust levels under all flight conditions, and completion of missions within the time allocated.

Propulsion Systems for UCAVs

So far, discussion of propulsion systems was limited to electronic drones, microdrones, and conventional small UAVs for commercial applications. Now, propulsion systems and their performance requirements for large-size UAVs and UCAVs will be discussed with emphasis on weight, size, endurance capability, payload capacity, and flight control sensors aboard the UAV. Note that the UCAV equipped with a fuel cell–based propulsion system is expected to carry out multiple combat missions including intelligence gathering, reconnaissance, surveillance, target tracking, target acquisition, target illumination, and deployment of laser-guided Hellfire missile to eliminate the threat. A high-performance UCAV aircraft similar to Predator B, as shown in Figure 5.4, can be operated using the combination of fuel cell (Figure 5.5) and microjet with minimum RCS and acoustic noise. This type of fuel cell is capable of providing high energy density and high discharge rates and discharge durations ideal for medium-size UAVs and UCAVs, provided the RCS of the fuel cell meets the minimum weight, size, and safety requirements. It should be noted that the flight control of the UCAV aircraft in three-dimensional coordinate space (Figure 5.6) is in the hands of an experienced ground control operator. This particular UCAV is fully equipped with appropriate EO and EM sensors, satellite-based communication receiver, GPS receiver, high-speed computer, and other necessary sensors. UAV orientation and relevant motion variables including aircraft body axis and earth axis in the x–y plane along with the angle of attack (α), roll angle (θ), yaw angle (ψ), side slip angle (β), pitching moment (M), and axial force (X) are shown in Figure 4.8. The UAV motion is fully described by the axial force (X), normal force (Z), and pitching moment (M). Note that forces X, Z, and M are assumed to be dependent on the UAV aerodynamic parameters,

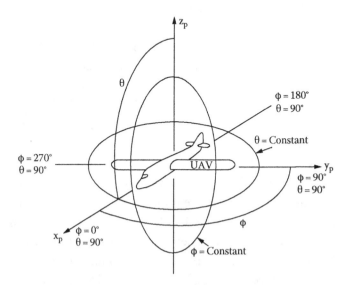

Figure 5.6   Typical UAV flight performance in three-dimensional coordinate space system.

propulsion power, and gravitational effects. Note that in simulation and testing of UAV or UCAV performance, longitudinal equations of motion are of critical importance. The ground control operator will handle the UAV aerodynamic flight characteristics regardless of propulsion system capability while the aircraft is flying.

The future generation of UAVs and UCAVs must explore the possibility of using $LiSOCl_2$ batteries and propellers to drive these vehicles in space. This propulsion concept offers the lowest RCS, minimum acoustic noise, absolutely minimum weight and size, and reliable power source. Battery research scientists feel that the high reliability and the 40-year longevity of such batteries would offer the right future for the next generation of small UAVs and micro-UAVs involved in border patrol, reconnaissance, and surveillance missions. In addition, the electrolyte used by these batteries offers higher thermal efficiency, improved thermal conductivity, and enhanced battery output energy at elevated temperatures with no compromise in safety and reliability.

Description of Various Propulsion Systems Currently Being Used by UAVs and UCAVs: Research studies undertaken by the author would reveal that various turbofans and turboprops are currently being used by UAVs and UCAVs. The research studies further reveal that a single Rolls-Royce turbofan system is used, where higher propulsion speed and platform weight close to 1100 lb are the principal requirements. It is extremely important to mention that gas turbines are mostly used by commercial power supply companies for the generation of high electrical power levels exceeding 1000–5000 MW and for the propulsion of ships or aircraft carriers. Note that steam power turbine generators are widely used by commercial utility companies in generating electrical power ranging from 1000 to 3500 MW. Deployment of micro gas turbines for UAVs or UCAVs is possible, provided the weight and size of the gas turbine and its components are within the specifications defined for the UAV or UCAV.

Comprehensive studies performed on propulsion systems for UAV and UCAV applications and jet engines with appropriate ratings have been deployed for the full-size UAVs and UCAVs. Both Russia and China have used the MIG-21 jet engines in their UAVs. Note that MIG-21 jet engines suffer from reliability and maintenance problems.

Israel has deployed a propulsion system using gasoline-based or kerosene-based two-stroke ICEs in their UAVs, which are exclusively developed for export purposes. Third-world countries are customers for the Israeli UAVs, because of minimum cost and complexity. These UAVs are very cost-effective and reliable and are best suited for undertaking intelligence gathering, reconnaissance, surveillance, and target tracking missions. This country has designed and developed small UCAVs exclusively for conducting RSTA missions.

Description of Propulsion Systems Deployed by UAVs: This section will describe briefly the propulsion systems deployed by various UAV programs. Propulsion systems using two-stroke ICEs will not be included in this section, but emphasis will be placed on propulsion systems deployed by various UAV programs.

Turbojet engines (WP-13) are capable of providing thrust levels greater than 9600 lb and are deployed by Soar Eagle and Global Hawk UAVs. Note that these UAVs are best suited for undertaking high-altitude, long-endurance (HALE) missions. Advanced turbojet engines with higher thrust capability could be deployed by U.S. high-performance UAVs such as Predator models B, MQ-1, or RQ-1.

A 100 HP, rear-mounted, turboprop engine is used in UAV configuration to drive a three-bladed propeller in a "pushing configuration." Rear mount offers much lower RCS as seen by the enemy radar and significantly reduced IR signature as seen by an IR missile head.

American Lycoming four-stroke ICEs with 150 HP rating have been deployed by some UAVs engaged in IRS missions with endurance durations ranging from 12 to 24 h.

Summary

Chapter 5 is dedicated to propulsion systems for electronic drones, tactical drones, micro-UAVs, large UAVs, and UCAV applications. Research studies performed by the author on various types of propulsion concepts seem to reveal that high-capacity Ni–Cd batteries are best suited for commercial drones, tactical drones, and quadcopters because they offer low cost, high reliability, zero maintenance, and minimum weight and size. As a rule of thumb, high-capacity propulsion systems involving jet engines or turbofan engines may be required to meet the requirements of UAV and UCAV platforms designed to carry out long-duration, complex military missions. Such platforms must operate at high altitudes greater than 50,000 ft or so to avoid IR missile attack and detection by enemy radar while the UAV platforms are engaged in long-endurance and complex military missions. In essence, the selection of a specific propulsion system must provide an iron-clad guarantee for high reliability of the platform, security of the vehicle, enemy radar detection, and safe return of the UAV to its base after completion of assigned missions. Note that safety of the UCAV platform is of paramount importance, because the vehicle is fully equipped with deadly laser-guided Hellfire missiles, high-performance EO and EM sensors, forward-looking infrared system, high-resolution side-looking radar, surveillance and reconnaissance receivers, communication receiver, GPS receiver, high-speed computer, and a host of high-resolution IR and optical cameras.

Note that propulsion sources for commercial quadcopters and electronic drones must be selected based on brief trade-off studies with particular emphasis on cost, reliability, weight, and size. Cost and reliability must be given serious consideration for the components of airborne systems such as UAVs. For the aforementioned applications, rechargeable Ni–Cd batteries are best suited because of high reliability, consistent performance, minimum weight and size, gasoline-based charging capability during the flight, and short-duration endurance performance capabilities.

Recently, a company has developed an 18 in. wide hexacopter equipped with tethered concept that eliminates the need for batteries. The fiber-optic cable deployed by the tether provides communication and electrical energy for the hexacopter.

Recently, tactical drones have been developed using a hybrid propulsion system technique, which involves a gasoline-based ICE to charge the Ni-Cd batteries during the vehicle flight. As mentioned before, the propulsion system for the micro-UAVs or electronic drones could use high-capacity batteries, $LiSOCl_2$ batteries, or fuel cells as long these components meet the stringent weight and size requirements. In the case of batteries, a number of critical parameters must be considered such as reliability, consistent performance under extreme temperature environments, and acceptable battery performance at a cryogenic temperature close to 80 K. Such cryogenic operating temperatures can be expected at altitudes greater than 50,000 ft.

The Dragon Eye UAV designed and developed by Naval Research Laboratory (NRL) uses a Ni-Cd battery pack weighing only 5 lb with minimum package dimensions. This micro-UAV has demonstrated a non-stop flight for a period exceeding 50 min over a range exceeding 40 mi, while traveling at an altitude close to 10,000 ft. Because of unique stealth feature, the UAV can be most ideal to carry out intelligence collection, reconnaissance and surveillance missions over hostile lands.

Research studies undertaken by the author on fuel cells seem to indicate that IEM fuel cells could meet the propulsion system requirements for heavy-duty tactical drones equipped with multiple EO and EM components and are engaged in carrying out multiple missions. Note that such fuel cells have demonstrated power ratings ranging from 3 to 5 kW, which can be adequate for completion of tactical missions ranging from 3 to 4 h.

LiSOCl$_2$ batteries can play a critical role in a propulsion system for a heavy-duty tactical drone or a medium-size UAV because of their proven reliability, extremely long service life exceeding 40 years, and consistent performance under extreme temperature environments. Note that lithium is a nongaseous metal that offers high specific energy (high electrical energy per unit weight) and maximum energy density (or maximum energy per unit volume) among all available battery chemistries. These batteries have demonstrated consistent performance over the $-55°C$ to $+125°C$ temperature range and demonstrated a reliable performance even at a cryogenic temperature of $-80°C$.

Since quadcopters typically operate at lower altitudes not exceeding 500 ft, gasoline-based two-stroke ICEs will be found most suitable as well as reasonably economical. However, to avoid enemy detection based on black exhaust smoke, it will be most appropriate to deploy rechargeable Ni–Cd batteries in quadcopters. Tactical drones equipped with a hybrid propulsion technique could yield long-duration endurance capability with minimum cost and complexity, with no compromise in vehicle reliability or safety.

For medium-size UAVs or full-size UAVs, there is a wide variety of propulsion systems ranging from turbojet engine to turbofan engine to four-stroke, high compression ratio, ICE. Published technical papers indicate that the U.S.-made Predator UAV models B, MQ-1, and RQ-1 could deploy either turbojet or turbofan as a propulsion system, because both have enough capacity to power the EO and EM systems and their accessories. It is justified to state that both engines could meet the propulsion system requirements for Predator MQ-1 and RQ-1 models. Microjet engines are available to meet lower propulsion power requirements.

Russian and Chinese UAVs deploy the turbojet engine as a propulsion system. This propulsion system is currently used by the Indian Air Force MIG-21 and SU-27 jet interceptors. However, this particular propulsion system suffers from serious mechanical, reliability, and maintenance problems. The Indian Air Force claimed a loss of 107 MIG-21 jet fighters, SU-15 interceptors, and a loss of undisclosed number of best pilots, because of the aforementioned problems.

British Aviation (BAe) has designed the UCAV Mentis, which weighs around 11,000 lb and is powered by a Rolls-Royce turbofan engine. This UCAV is specially designed to carry out high-endurance, long-range combat missions. France has designed, developed, and tested UAVs for combat-related activities.

Countries including India, Israel, Brazil, South Korea, South Africa, and some other nations are deeply involved in the design and development of UAV platforms. However, intelligence gathering, reconnaissance, and surveillance are the major functions of these UAVs.

References

1. P. O'Shea, UAVs make big slash into commercial spaces, *Defense Electronics*, 15–18, May 2015.
2. A.R. Jha, *Next Generation of Batteries and Fuel Cells for Commercial, Military and Space Applications*, CRC Press, Boca Raton, FL, 2012, pp. 112–113, 259–261.
3. Editor, Tether and drones provide an extra pair of eyes, *IEEE Spectrum*, 27, February 2015, pp. 15–17.
4. S. Jacobs, *NASA Tech Briefs*, May 2011, pp. 18–19.
5. Press Release, GE aviation and Woodward combine fuel system expertise for new joint venture, General Electric, Cincinnati and Woodward Global Company, Fort Collins, CO.

6

UNMANNED AUTONOMOUS VEHICLE TECHNOLOGY

Introduction

This chapter exclusively deals with the autonomous vehicle technology and sensor requirements for unmanned aerial vehicles (UAVs). An unmanned vehicle (UV) will have no operator aboard the vehicle and will have all the necessary sensors and high-speed computers essential to undertake the mission tasks assigned to it. Of course, this vehicle will be in constant touch with the ground-based operator through communications. Under emergency conditions, the ground-based operator will have the authority to extend or abort a specific tactical mission in order to save the costly, complex, secret vehicle from falling into enemy hands.

Tactical experts and military authorities predict that tactical success in future military conflicts will depend upon unmanned autonomous vehicle performance. Research studies undertaken by the author on such vehicles and their tactical roles seem to reveal that stringent weight, size, and power requirements should be given serious consideration to minimize fuel consumption and to extend endurance and range capabilities, if needed, under combat environments. State-of-the-art electro-optical (EO) and latest electromagnetic (EM) sensors and high-speed computers will play critical roles in assisting unmanned autonomous vehicles in executing tactical missions. Since this vehicle is designed to operate autonomously, its structural integrity, platform operational reliability, and consistent operations of sensors and computers aboard are of critical importance. The ground control operator will have hands-off policy under normal operating conditions of the unmanned autonomous vehicle. However, the ground control operator will be available if and when needed. Furthermore, this ground control operator will always be in touch with the vehicle while in flight. Note that the global positional system (GPS) and satellite communication receivers will be available at all times to the vehicle and the ground control operator will maintain the UAV's autonomous capability.

Example of UAV with Autonomous Capability

According to a recent publication [1], the Navy's X-47 B unmanned combat air vehicle (UCAV) system became the world's first pilotless, semiautonomous aircraft, which has successfully demonstrated takeoff and landing from the carrier deck all by itself. Note that automatic takeoff and landing operations from the ground are more easy and safe compared to those operations from a carrier deck. This particular aircraft was built by Boeing Company in 2011. This is not 100% autonomous, but its fully autonomous version is expected to be ready in the near future.

This X-47 B aircraft is 38 ft long and is controlled by the carrier deck operator or ground operator using keyboard commands and mouse clicks. The human operator from the carrier deck tells the aircraft to go to a certain point on the map, and the X-47 B vehicle complies with the command. This is pretty close to an unmanned aerial autonomous vehicle as far as the author is concerned. It can fly at subsonic speed with a range exceeding 2000 nm. This aircraft has demonstrated in-flight refueling capability all by itself, with a bit of help from a tanker aircraft. According to reliable sources, the next generation of X-47 B aircraft, as shown in Figure 6.1, is expected in the near future and will need a lot of situational awareness for 100% autonomous capability.

The Predator and Reaper military drones essentially are remotely piloted vehicles (RPVs), which are controlled by human operators sitting in ground control stations thousands of miles away from the combat zones using control sticks. Note that combat drones or UAVs or RPVs are not autonomous or robotic platforms, and they are controlled by human operators in real time. It can be stated that autonomous capability will drastically change aircraft carrier operations.

Figure 6.1   X-47 B UAV with quasi-autonomous capability.

Encouraging Signs of Autonomous Capability in the Auto Industry

UAVs have become the essential component of the modern battle-field plan. Quasi-automatic landing and takeoff capabilities have been demonstrated by the Predator B and MQ-9 remote piloted vehicles (RPVs). However, these landings involved the automatic application of engine reverse thrust and brakes to bring the aircraft to a complete stop. More experimental activities are required to demonstrate the autonomous capability of the UAV.

Recently, research scientists and automobile design engineers have been actively involved in the design and development of autonomous cars. These engineers predict that autonomous vehicles with no drivers may come sooner. It is estimated that a smart taxicab could take passengers safely to their destinations. It is most likely that these smart automobiles or cabs may replace on-demand ridesharing commercial services such as Uber and Lyft. The foregoing autonomous car is a typical example. Since autonomous cars here are more or less using high-speed computer technology, smart components, and smart materials, the author predicts that there would be no further delay in the existence of an autonomous UCAV.

Smart Materials for UAVs

Rapid progress of autonomous technology strictly depends on smart sensors and materials. Comprehensive studies undertaken by the author seem to reveal that smart materials and advanced component technology are most desirable to achieve optimum vehicle performance, high reliability, and structural integrity of the vehicle under adverse aerodynamic and climatic conditions. The studies further reveal that the drones operated by military organizations or UCAVs operated by field commanders will benefit the most from the availability of smart materials.

Highlights of Smart Material Properties: Tensile strength, flexural strength, and stiffness are the major material characteristics of smart materials. Research studies undertaken by the author on smart materials reveal that temperature stress and magnetic field can produce a material change in shape, strength, and stiffness. This means that a smart material can move without requiring a mechanical drive source or a motor or a hydraulic device, leading to significant reduction of weight and size in the vehicle. Research studies further reveal that smart materials offer optimum stiffness close to 270 MPa or 38,200 lb/in^2 Note that high stiffness is of critical importance for the development of a combat vehicle. The major benefits of smart materials can be summarized as follows:

- A smart material does not require a mechanical moving device.
- High stiffness at elevated temperatures.
- High mechanical integrity of the parts made from smart materials.
- Smart materials offer a wire brake release function with minimum cost and complexity.
- These materials offer a significant reduction in weight, size, and labor cost for the parts, which will significantly increase the endurance capability of the vehicle leading to execute several combat-related missions.
- Smart materials have demonstrated several unique properties that can change in a controlled fashion by external stimuli such as stress, temperature, moisture, and electric and magnetic fields.

Smart Components for UAVs

Smart and microcomponents are most ideal as smart materials for UAV applications, where minimum weight, size, and high reliability are the principal requirements. Under this subtitle, only widely deployed critical components for UAV applications will be briefly described with emphasis on weight, size, reliability, and overall performance of the component under combat environments. Such microcomponents including gyros, accelerometers, motion controllers, and fluidic actuators will be summarized with emphasis on superior performance, improved reliability, and significant reduction in weight, size, and power consumption. The parameters mentioned will improve the endurance capability and range performance of the UAV platform.

Gyros for UAV Applications

Recently, nanotechnology-based microsensors have been designed and developed to monitor harmful environmental parameters under battlefield conditions for the safety of soldiers [2]. Microelectromechanical system (MEMS)-based gyros [2] are best suited for UAVs to meet precise and safe navigation, smart munitions, robotic capabilities, airborne tracking radars, and a host of other military systems. Precision navigation aids vital for safe and reliable navigation, smart munitions, and battle carriers are widely deployed in military applications, robotic operations, airborne radars, and other military sensors. Note that the highest-precision inertial measurement units (IMUs), such as ring-laser gyros and fiber-optic gyros, have been in use for several years, though these devices provide reasonably good performance, but these suffer from high cost, large size, and excessive power consumption. MEMS engineers have developed MEMS-based gyros that yield better performance with minimum weight, size, cost, and power consumption. The users of these gyros reveal that these devices provide precision guidance for military, commercial, and aerospace vehicles.

The critical components of a MEMS-based gyro include various functional elements, self-test port, power management circuit, digital processing network, signal conversion circuit, and calibration circuit as shown in Figure 6.2. It is interesting to point out that these devices are best suited for automobile and UAV guidance systems for vehicle

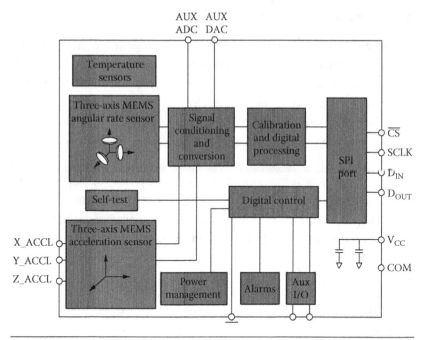

Figure 6.2 Block diagram of a MEMS-based gyro showing various interface and functional elements of the device.

stability control and precision guidance. Combining of MEMS gyros with accelerometers offers a backup means for the GPS. In this way, this backup technique provides significantly enhanced reliability for the automobile and unmanned aerial vehicles needed for adverse control environments. Note that gyro stability is considered the most critical performance parameter for UAVs. These gyros are capable of providing the moving platform stability in few tenths of a degree per hour. Note that this gyro has inherent biases and sensitivities to temperature, supply voltage fluctuation, and moving body vibrations. To control these sensitivities, self-test and calibration circuits are provided as illustrated in Figure 6.2. Note that short-term stability can be accomplished with minimum circuit components. However, long-term stability may require sophisticated filter networks and digital circuits.

MEMS-Based Accelerometers for UAVs: Precision accelerometers are required for commercial, industrial, and aerospace system applications. Low-cost accelerometers are widely used for consumer electronic applications. MEMS accelerometers are best suited for unmanned aerial vehicle applications. Studies undertaken on accelerometers seem to reveal that low-gravity sensors are most ideal for electronic systems that require the detection of small changes in the force resulting from motion, tilt, shock, vibration, and positioning. These research studies further indicate that the two-axis accelerometer is capable of sensing in both lateral and perpendicular planes. However, the two-axis accelerometer known as XZ-axis accelerometer is considered the most effective device.

An accelerometer with various sensitivities is available to meet the needs for multiple applications and functions. MEMS-based accelerometers are best suited for embedded systems and GPS sensors. Key features of MEMS-based accelerometers can be summarized as follows:

- Full range of directional motion
- Low-G capability
- Minimum current consumption (500 µA)
- Sleep-mode design feature with 3-µA consumption
- Low-voltage operation ranging from 2.2 to 3.6 V
- Fast power-up response less than 0.001 s
- Microelectronic interface compatibility

Fluidic Actuators: A fluidic actuator with no moving parts has been recently developed to control fluid flow, which will ultimately result in significant performance of the moving parts of the UAVs and improved fuel efficiency, leading to significant improvement in the endurance and operating range of the UAV. In brief, the fluidic actuator technology will improve the overall performance of the UCAV.

It is interesting to point out that the flow control actuators, also known as fluidic oscillators or sweeping jet oscillators use the Coandă effect to generate spatially oscillating jets. Note that the fluidic control actuators can be embedded directly into a control surface such as a wing or a turbine blade in order to reduce the flow separation, increase the aerodynamic lift force, reduce aerodynamic drag, and improve heat transfer capability. Major advantages of fluidic actuators can be described as follows:

- High reliability because of absence of moving parts
- Significantly improved aircraft performance
- Reduction in aerodynamic drag

- Improvement in aerodynamic lift
- Significant reduction in weight and size of the actuator devices
- Can be embedded directly into the control surfaces
- Improvement in fuel efficiency
- Significant reduction in fuel consumption
- Offers longer endurance capability of the UAV
- Enhanced heat transfer efficiency
- A minimum of 60% improvement in aerodynamic performance of the UAV
- Effectively decouple the oscillations from the amplitude variations
- Offer high mass flow rates
- Yield simple and compact architectural features
- Do not require additional equipments such as oscillations, synchronization, and decoupling
- Maintenance-free operation
- Highly scalable actuator design
- Can be manufactured using different materials and still can be used in harsh environments

Motion Controllers for UAV Application

Motion controllers are widely used in controlling two- or three-degree axes with utmost precision. The core axis connections can be configured in pulse and direction outputs to drives or incremental encoder feedback inputs or simulated encoder outputs. The three-axis variant allows one axis to be set up as an extended axis where the connection may be configured as an input if required. Important features and performance capabilities of motion controllers can be summarized as follows:

- A 64-bit efficient processor.
- Digital and analog input/output count.
- Built-in Ethernet.
- Ability to add robot transformations or synchronize motion with double-precision floating point.
- Controller ability to undertake linear, circular, helical, and spherical interpolations across all axes as well as flexible CAM-shaped and linked motion with remarkable accuracy.

- The controller offers eight 24-VDC digital inputs including six 20-µs registration inputs and four 24-VDC bidirectional input/outputs.
- Two 121-bit analog inputs are available.
- Ethernet port is provided for programming and connections for other devices if required.

Smart Materials for UAV Components [3]: Use of smart materials for certain UAV components and devices will provide higher reliability, improved thermal conductivity, and enhanced mechanical integrity under harsh mechanical and thermal environments. Research studies performed by the author reveal that UAV components and devices made from smart materials offer the highest mechanical strength under adverse thermal and rough atmospheric environments. In addition, no other materials can match the tensile, compressive, and flexural strength of smart materials. Unique characteristics of smart materials and their potential applications can be briefly summarized as follows:

- Pd–Ag smart alloy material is widely used in the fabrication of microfabricated membranes, where strength is the principal design requirement.
- This particular smart material (Pd–Ag) yields maximum transmembrane pressure.
- Its yield strength is close to 80 MPa and Young's modulus exceeds 150 GPa.
- This alloy has demonstrated a remarkable rupture capability.

Military Role of Unmanned Autonomous Vehicle [4]

This chapter deals with the autonomous design of unmanned aerial vehicles that will change the course of future military conflicts. The proposed autonomous aircraft architecture will play a critical role in military conflicts with no pilot in the fighter aircraft. For this particular unmanned aircraft, computer, sensor, and display requirements will not only be difficult but also very complex when so many different sensors and systems will be crowded in to a UAV. This vehicle will be managed by the ground control operator, who has comprehensive experience as a seasoned fighter/bomber pilot and is sitting in an air-conditioned room thousands of miles from the point of military activity. The ground control operator will be always in touch with the UAV through secured communication and he can give command to a computer for any change in execution of a mission or to abort any mission task. In addition, the ground operator can give a command to a particular sensor to look for targets in a specific area and direction. In brief, the ground control operator has full authority and responsibility for all sensors aboard the vehicle.

Role of Electronic Switch Modules

Electronic switch modules operate in conjunction with VFX displays or as standalone interface devices. These electronic switch modules assist the ground operators to get the most out of their systems or machine performance, while allowing OEMs to drastically simplify CAB wiring layouts.

PRO-FX Application Software: Powerful software has been developed by Eaton Corporation (NY) to deal with UAV simulation. Eaton has introduced the most advanced programming, commissioning, and diagnostic software suite for easy integration of machine functions with the electronic control system.

Power Insight 2.0 Software Capability: According to the software designer, this software provides fast, accurate, and side-by-side comparisons of UAV designs, which test engineers or the program managers can evaluate in a single common test and evaluation environment.

dSPACE SCALEXIO Software: Note that the patent version of dSPACE SCALEXIO software and its hardware in the loop provides additional functionality, supporting SAE J2716 single-edge nibble transmission (SENT) 2010 and Ethernet protocols. The unique properties of this software are identified as follows:

1. Offers flexible and scalable failure simulation
2. Provides SAE Standard J 2716 SENT 2010, SENT in and SENT out to simulate and capture communications data for ECU software validation and testing

Eaton's array of electronic components and software are known for their reliable and efficient system design and integration capability.

Software that supports a flying vehicle equipped with navigation and obstacle avoidance sensor is described under three distinct scenarios as follows:

Scenario one: This scenario describes vehicle performance including reliability and safety, which is equipped with a navigation system and obstacle sensor. The unmanned autonomous vehicle is flown toward a designation point. The navigation system aboard the autonomous vehicle is capable of detecting any obstacle in the path or any flying object and can avoid collision with the aircraft in the flying corridor.

Scenario two: This scenario particularly concentrates on the vehicle's maneuvering capability. It is assumed that the autonomous vehicle is flying under normal atmospheric environments, and an open-loop change maneuver is performed at a specified speed to demonstrate the realistic nature of the vehicle dynamics.

Scenario three: This scenario focuses on the vehicle simulation capability that can demonstrate the high-fidelity, multibody dynamics, EO and EM sensors, actuators, control and navigation capability in difficult atmospheric environments while it is travelling at various speeds that are assumed to be useful or appropriate for the autonomous vehicle performance analysis and vehicle aerodynamic performance verification purposes.

Role of Critical Miscellaneous Components

This section will briefly describe the important roles of various critical miscellaneous components and sensors aboard the autonomous vehicle.

Role of Programmable Controllers: Eaton programmable controllers take control of all system functions of a dump truck, crane, or other mobile platform on highway equipment.

Role of VFX Programmable Displays: Eaton VFX programmable displays are robust and rugged for airborne or ground moving system applications. Its optically bonded LCD display assembly provides remarkable clear viewability.

Integrated Simulation Capability of UAV

Integrated simulation capability for UAV platforms requires real-time, high-fidelity dynamics with control sensors and environmental models in the closed loop for a representative autonomous vehicle [5]. Note that the simulator's architecture will present multiple selections of different fidelity levels as well as model parameters across the full modeling spectrum. It is desirable to mention that the simulator's architecture will also provide an analysis scenario in a modular way to swap EO or EM component models for changing vehicle, control, and sensor behavior.

The UAV is treated like a suspension system consisting of a large number of distinct bodies or vehicle components that are contained within a single kinetic loop. Because of this, these suspension elements or components have a large number of internal degrees of freedom. However, due to the constraints imposed by these components or subsystems such as sensors to a vehicle frame or chassis, they have a single independent degree of freedom.

It can be further stated that the suspension modeling based on three algorithmic techniques can be tested and benchmarked to solve the multibody dynamics of the suspension system, which can be recognized as the unmanned autonomous vehicle. This vehicle simulation should be simulated in only one operation known as atmospheric environment. A graphical representation of this environment can be achieved using a variety of tools, and ultimately, a digital elevation map can be extracted from it. Note that a digital elevation map or horizontal map can be created using this technique and imported into the computer simulation. This simulation requires the following three distinct scenarios to be performed:

Scenario no. 1: This particular scenario supports the real-time use of an Ethernet protocol that enables sensor devices to communicate with SCALEXIO's processing unit.

Scenario no. 2: This scenario facilitates input/output functions to convert into graphical configuration.

Scenario no. 3: This scenario demonstrates failure simulation capability using a single hardware license. This simulation can be activated by the existing input/output channels. This scenario also allows to compare test systems. Note that both the systems and networks are built with common hardware components, which allows the use of subsystems of a network system for component test and performance evaluation.

Note: The latest software including SYNECT, SCALEXIO, and VEOS, which are vital for complex system simulation activities, are available from Eaton Corporation.

Supervisory Control of UAVs [6]: So far more or less all UAVs or UCAVs are operated and managed by the ground control operator sitting in a room located thousands of miles away from the airborne vehicle. Regardless of where the operator is sitting, to undertake the supervisory control of an airborne vehicle is very difficult and complex and requires a sharp and very intelligent control operator with better flying experience than a fighter/bomber pilot. The operator must have comprehensive, multiyear flying experience and must have demonstrated the handling capabilities of the sensors aboard the vehicle, and be capable of quick decisions without outside assistance. In brief, the supervisory control of an autonomous vehicle requires a seasoned pilot with multiyear jet pilot experience and must have demonstrated thorough knowledge of the sensors aboard the autonomous vehicle.

Quantitative Supervisory Model for UAV: The development of a supervisory quantitative UAV model is not only very complex but also extremely difficult [6]. The model designer must have adequate knowledge of the performance capabilities and limitations of various EO and EM sensors aboard the autonomous vehicle. This model development requires a task-centric approach to interface a workable design, which addresses an explicit undertaking of certain tasks or actions that need to be performed by the vehicle operator in a specified time frame. Essentially, the representation of activity in terms of a specific task serves as a trace cursor in the system that allows the vehicle design engineers to monitor the workload as well as the tasks performed by the team members in a specified duration. Research scientists predict that in the supervisory vehicle control, the focus must be on the flow of tasks to be performed by a system composed of human task performers and automated servers such as software agents.

It is assumed that the quantitative models are fully capable of analyzing the dynamic systems of flow that have been developed in the domain of queuing theory. This quantitative approach can be extended to include the supervisory control of UAVs. In this analysis, the "customers" in the queuing are the sequenced tasks, which must be completed by the human software agents, high-speed powerful computers, and UVs [7].

Besides the physical platform, autonomous vehicle agents working as team members are the prevalent tools to accomplish the mission assignments. Remember that the coordination of actions and interactions among the UAV systems, manned systems or sensors, and a command group is essential to accomplish future mission goals. This creates a new problem and must be addressed carefully and quickly. The real problem is "how to maintain an adequate workload to avoid information workload and ultimately net loss of situational awareness."

Leading MIT scientists have developed a technique known as Research Environment for Supervisory Control of Heterogeneous Unmanned Vehicle (RESCHUV) technique to address the UAV problem. It is interesting to point out that the RESCHUV approach was employed to test the supervisory control capability involving a task to evaluate the surveillance, reconnaissance, and identification (SRI) mission functions. Later on, the simulation was modified to undertake the following three distinct tasks:

Task 1: This task is dedicated to a complex mission scenario with an asset to protect multiple and simultaneous attacks by multiple airborne vehicles.

Task 2: This task involves a highly automated system such as a mission of definition of language.

Task 3: This task assumes a highly heterogeneous team consisting of at least three different types of UVs. The new version of this scenario is known as RESCHU-SP. It is interesting to note that in the RESCHU-SP scenario, a single operator manages and controls a team of unmanned aerial vehicles (UAVs), unmanned surface vehicles (USVs), and unmanned underwater vehicles (UUWVs) [6]. Now the supervisor or operator job is more cumbersome and complicated. He has to deploy more UAVs to identify and destroy the enemy aerial vehicles that are destroying important assets such as oil platforms and military installations. In addition, to defend the

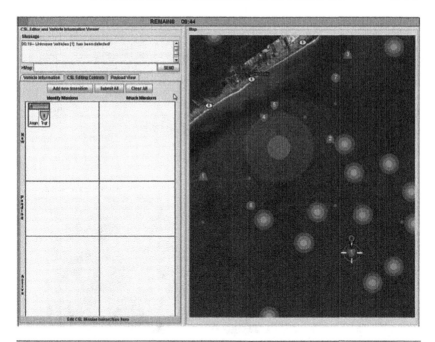

Figure 6.3 Interface for the RESCHU-SP simulations and the operator's display consisting of two main windows placed side by side as shown on the left. (From DiVita, J. et al., *A Queuing Model for Supervisory Control of Unmanned Autonomous Vehicles*, Space and Naval Warfare Systems Center Pacific, San Diego, CA, September 2013.)

costly installation or oil platform, the operator must divert the UAVs away from the hazardous regions that could damage UVs. The interface for the simulation can be seen in Figure 6.3.

The operator display shown in Figure 6.3 has two main windows placed side by side [6]. In one window, a geosituational display depicts the spatial position of the assets and the oil rig, unknown contacts, enemy contacts, and hazard regions. The second adjacent window is a three-tabbed pane window that contains the following contents:

1. Vehicle information display
2. Collaborative sensing language (CSL) editing controls
3. Payload window as shown in Figure 6.4

The operator must engage in five distinct tasks:

1. Assign a competent person to identify the unknown contact.
2. Engage to identify an unknown contact.

Figure 6.4 Details of the payload view that allows an operator to identify visually the contact as friend or foe. (From DiVita, J. et al., *A Queueing Model for Supervisory Control of Unmanned Autonomous Vehicles*, Space and Naval Warfare Systems Center Pacific, San Diego, CA, September 2013.)

3. Assign an attack mission.
4. Engage in attacking the enemy contact.
5. Carefully evaluate hazard avoidance.

According to military experts, only the UUWV can attack enemy contacts.

It is essential that the UAV should fly in undetermined flight patterns to avoid enemy detection and in this way the operator can change to avoid hazard regions. In order to handle the situation carefully, the operator must select or identify an unidentified contact and then assign an unmanned security vehicle (USV) to confirm the presence of hostile contact. Once the USV arrives at the prescribed destination, the operator may engage the USV to identify the contact. This particular action will bring in a picture in the payload view as illustrated in Figure 6.4, which permits the operator to visually identify the contact as a foe or friend.

Once the contact has been identified as an enemy, a contact should be assigned for a UUWV to attack it immediately. Note that this sequence of actions to attack the enemy leading to complete destruction is analogous to the identification process. Ultimately, an order must be given by the operator for a UUWV to attack. Once the UUWV has arrived at the target location, the enemy icon flashes and the operator can select the icon, which allows to eliminate the target at full scale.

Upon elimination of the target from the geosituational display, it will confirm that the enemy is destroyed.

The close relationship between the operator, the CSL, and the three UVs and the manner in which they performed the assigned tasks can be modeled as not a work of interactive queues. It is highly essential to remind that in an open queuing system, the tasks arrive at each of the server locations. It should be noted that some tasks are processed and leave the system, but other tasks may be in the process of leaving from one server to another server. Under such condition, tasks can arrive at different queues and be waited upon by the different servers and return to the previous server before leaving the system. This confirms that the queuing theory provides the most effective quantitative tools to analyze the flow of tasks to and from each server. Furthermore, the queuing theory allows complete network analysis to achieve overall system performance. It should be pointed out that the EO and EM components and devices involved in formulating and analyzing a queuing system can be extracted from the RESCHU-SP scenario.

It is the responsibility of the human operator to monitor, carefully watch, and manipulate the activities of multiple UAVs, including the situation assessment, decision-making, task planning, and corrective actions if required. UAV scientists believe that military authorities seek to enable agile and adaptive mission management and control for a mission-orientated team composed of UAVs, unattended ground sensors (UGSs), disengaged war fighters with mobile control stations, and an operator located in a central control station (CCS).

With UAVs equipped and authorized to replan and act without human input or control, the challenge is to develop methods for a human operator to effectively monitor the activities, which may include goal-directed task selection, situation assessment, decision-making planning, and future actions needed. Recommendations suggested by the CCS operator can be summarized as follows:

- Effective monitoring of the tasks
- Review of critical activities
- Manipulation of the assigned UAV activities
- Monitoring of the progress in the goal-directed task
- Situation assessment
- Decision-making planning
- Future actions needed

Military planning authorities believe that there is a challenge to present the correct information in a manner that provides quick and most accurate assessment of the multivehicle system engagement. The goal of this activity is to provide design concepts that can fully support the use of symbols and patterns capable of supporting "at-a-glance" recognition of complex military activities. Hierarchical pattern-oriented state diagram concepts have been developed to represent the autonomous activities of the vehicle. Whereas, layered finite-state machine diagrams combine with a control-timeline and payload inspection display, collectively referred to as *layered*.

It is highly desirable to point out that hierarchical pattern-oriented state diagram concepts have been developed to represent the autonomous activities of unmanned aerial vehicles.

Laboratory-based pattern recognizable interfaces for state machines (L-PRISM) will be integrated with the multi-UAV control system's tactical situation map (TSM) and system states to assist the operator to be aware not only of the vehicle's locations and their planed routes but also their mission goals, associated tasks, and states to achieve their future mission goals. Note that the L-PRISM concept, as illustrated in Figure 6.5, is essentially composed of three distinct display components:

1. The state diagram
2. The timeline
3. The payload viewer

The state diagram uses the conventions of finite-state machine diagrams, as shown in Figure 6.5, to represent the autonomous system

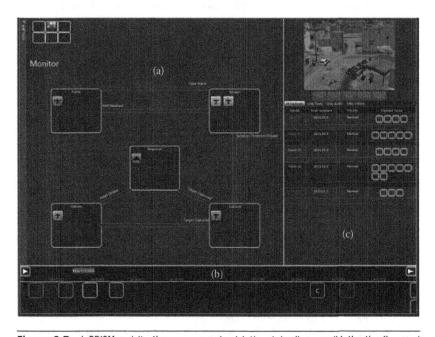

Figure 6.5 L-PRISM and its three components: (a) the state diagram, (b) the timeline, and (c) the payload viewer. (From Patzek, M. et al., Supervisory control state diagrams to depict autonomous activity, Air Force Research Laboratory, Wright-Patterson AFB, OH, June 6, 2013.)

activity to the operator in real time. In addition, the state diagrams identify the autonomy aspects through the state modes and transition arcs of the diagrams. Note that the state nodes display the elements that present the autonomous assigned tasks, showing what the vehicle is doing. It is equally important to note that the transition arcs represent the specific criteria for changing the tasks, showing why the vehicle should move to a new task. Multiple vehicles can be displayed in one diagram to accommodate multi-UAV monitoring and control. The state diagram shows simultaneously the state of each UAV participating in the mission through the vehicle icon. Note that the vehicle icons can show the vehicle type, identify the vehicle color by its call sign, and provide a time-on-task clock. It is desirable to mention that hierarchical pattern-orientated state diagram concepts have been developed to represent the autonomous activities of unmanned aerial vehicles.

Autonomous vehicle designers believe that an autonomous system can only make task and subtask changes that essentially follow the transition arcs. For tasks that have no transition arcs, task changes can only be made by the operator. In general, the vehicle control within the L-PRISM has the flexibility to support different levels of automation, as long as there is an adequate communication with the particular vehicle or vehicles.

It should be remembered that the L-PRISM concept uses a layered arrangement of tested state diagrams, which provide representation of autonomous tasks at varying levels of abstraction. These levels of abstraction are expected to improve the understanding of and management of autonomous activities in part by displaying connections between the actions, plans, and goals.

It should be noted that the timeline presents the vital information on significant mission activities and provides effective controls to review that information provided by the timeline across the control station in a timely order. Note that an event is represented by a colored tile containing a letter or icon that indicates the vehicle change with task change or payload delivery. It should be remembered that the colors match the vehicle color, which is used throughout the ground control station, and icons identify the type of payloads. The ground station controllers can proceed forward or backward in the designated time duration to investigate the tasks and actions of the vehicles assigned and the rationale behind those actions.

The payload viewer shown in Figure 6.4 displays a selected list of mission-related data referred to as payloads. It is interesting to point out that typical payloads are the images or videos from UAV EO and EM sensors but could be operator-generated images or videos. Remember that all operators, which can be dismounted war fighters, the stationary fighters with mobile control stations, and the stationary CCS or ground control station, can generate and send text messages or "voice notes" and audio recordings as payloads.

Research studies performed by the author seem to indicate that the L-PRISM concept described here is an evolving supervisory control display concept with vital information for the supervisor. This supervisory control display shows a great promise for providing important attributes and features for displaying vital information from a highly autonomous multivehicle system.

In summary, the L-PRISM method will allow the human operator sitting in a ground station to monitor, inspect, evaluate, and manipulate the activities of multiple unmanned aerial vehicles in real time, including real-time situation assessment, tactical decision-making capability, strategic planning, and appropriate actions needed to control the battlefield conflicts. Finally, it is of critical importance to emphasize that the human pilot or ground control operator for this particular activity must be a seasoned human pilot with multiyear fighter–bomber flying experience, technically competent, familiar with the performance capability of the sensor aboard the UAV, weapon handling and deployment ability, and be quickly decisive, and absolutely fit physically and mentally.

Description and Performance of Sensors aboard Autonomous UAVs

This section deals strictly with the brief description and performance capabilities of EO and EM sensors aboard UAVs. It is essential to emphasize that the sensors aboard the UAV must have state-of-the art performance capabilities and have satisfied the performance specifications including high reliability and safety under harsh thermal, aerodynamic, and atmospheric environments.

Electro-Optical Sensors for Possible Applications in Autonomous Vehicles: The author will identify various EO sensors for possible applications in UAVs. Potential EO sensors include EO/infrared (IR) laser designator payload. These combined sensors provide surveillance, reconnaissance, intelligent, and target detection and acquisition functions irrespective of weather conditions. These sensors offer vital information on the targets under harsh battlefield environments. Most of these sensors operate in 2–3 μm regions. However, a CO_2 laser operating in the 10.4–10.6 μm range is deployed in target illumination for the laser-guided missiles. Note that such sensors must use technology that will yield minimum weight, size, and power consumption to achieve long-endurance capability and minimum fuel consumption. Note that such sensors have been used by U.S. Customs and Border Protection in drug interdiction with remarkable success.

Forward-Looking Infrared Sensors for Autonomous Vehicles: Military publications and technical literature will indicate that FLIR sensors play a critical role in battlefield conflicts. Research studies undertaken by the author on EO sensors reveal that the latest FLIR sensors with minimum weight, size, and power consumption are available, which are best suited for UAV applications. These sensors come with three fields of view (FOV), namely, narrow FOV with $10° \times 15°$ capability, medium FOV with $15° \times 25°$ capability, and wide FOV with $20° \times 35°$ capability, are available for UAV applications to conduct surveillance, reconnaissance, intelligence gathering, target detection, acquisition, and tracking missions. Mini-FLIRs with minimum weight, size, and power consumption are specially designed for autonomous vehicles and are available for UAV deployment. The mini-FLIRs provide precision target detection, identification, and tracking when operated in high-resolution, narrow FOV mode and are installed in the most appropriate locations on the UAV fuselage or nose section.

Infrared and Optical Cameras for UAVs: Preliminary research studies performed by the author on optimum sensor installation locations seem to indicate that a forward-mounted colored camera or IR camera will be found most helpful in flying the UAV aircraft at lower altitudes. The studies further indicate that the IR and TV cameras when installed at appropriate locations in the fuselage play critical roles in EO-guided missiles in combat environments.

A six-rotor UAV was recently designed for military application especially to demonstrate optimum UAV strike capability when equipped with high-resolution EO and IR sensors. This vehicle demonstrated fast response, strong defense penetration capability, high payload capability, and remarkable reliability under harsh operating environments. This vehicle also demonstrated successful and deep penetration, precision target acquisition, surveillance, damage assessment, and other combat-related missions.

Miscellaneous EO/EM/IR Sensors and Components Available for UAVs: The latest survey conducted by the author on EO/EM/IR sensors and components seems to reveal that these items have been developed especially for autonomous vehicles, which offer high reliability, minimum weight and size, improved accuracy, and high-resolution images. The following EO/IR sensors and components are best suited for unmanned autonomous vehicle components:

- Optical camera
- Optical video camera
- IR line scanner
- Multispectral camera
- Pod-mounted laser radar
- Pod-mounted laser illuminator
- Differential GPS up to a centimeter level in flight path and landing and allows 2 UAVs in one ground station
- Microwave and mm-wave surveillance, reconnaissance, and EW receivers capable of meeting the performance, weight and size, and power consumption specification requirements
- Compact and lightweight RF jammer if weight and size requirements are compatible with performance specifications
- Side-looking radar
- Forward-looking infrared (FLIR) sensor with multiple FOV capabilities

The UAV operator or a program manager will select the EO, EM, and IR sensors and components that will assist him in undertaking and meeting the combat mission requirements. All the sensors or components described here are available for deployment in UAVs, and the choice for the sensors will be made by military experts and vehicle program managers.

Propulsion Systems for Unmanned Autonomous Vehicles

The readers should be aware that propulsion systems have been described for small- (micro-UAV) and medium-size UVs in Chapter 5. Propulsion system requirements for autonomous vehicles will be more stringent compared to those for small- and medium-size unmanned aerial vehicles. Suitable propulsion systems are hard to find because of stringent requirements for the size, weight, endurance, fuel consumption, and jet engine reliability.

Brief Description of Some Current Propulsion System for Autonomous Vehicles: The following propulsion systems are available for unmanned autonomous vehicles, however the type of propulsion system is strictly dependent on the current autonomous vehicle design requirements and are approved by the engineering manager or program manager. Typical propulsion systems can be described as follows:

- High-power Tesla turbine engine
- Turbojet engine
- High-thrust turbojet engine
- Turbofan engine

Typical Performance Parameters of Propulsion Systems for Large UAVs: Preliminary research studies performed by the author on power plants for autonomous vehicles seem to indicate that jet engines in the turbofan and turbojet categories may be able to meet the propulsion system requirements for large-size vehicles such as MQ-1 Predator and MQ-1 Reaper. Enough engine thrust must be available keep the vehicle aloft at all altitudes under harsh atmospheric conditions and to meet the payload, endurance, and power consumption requirements for all the sensors aboard the vehicle. The following are the most critical power-related parameters for high-performance autonomous vehicles:

- Energy efficiency
- Calorific value of the fuel (BTU/lb of fuel)
- Power output-to-weight ratio
- Thrust-to-weight ratio
- Thrust lapse for jet engine
- Overall pressure ratio

Description of Propulsion Systems That Could Be
Deployed for Autonomous Vehicles

The author has investigated the performance specifications for some existing propulsion system designs that are currently used in quasi-autonomous vehicle designs. Propulsion systems with minimum weight and size and high reliability may not be readily available within the price range. Furthermore, in the author's opinion, it can be stated that two or three countries are capable of designing and developing such vehicles with stealthy features. The export of UAVs or technology capable of carrying 500 kg or 1100 lb payload over 300 km is restricted to many countries by the Missile Technology Control Regime (MTCR). However, the American military continued to undertake UAV research and development activities to conceive MQ-X Reaper and MQ-X Predator, which are very close in meeting the autonomous vehicle performance specification requirements. Note that the advanced versions of Predator and Reaper could represent full-size UAVs. These autonomous vehicles when equipped with suitable propulsion systems will not only meet the power consumption requirements for EO/EM/IR sensors but will also provide adequate thrust for the vehicle to operate under harsh atmospheric environments. Power requirements for the passive sensor components will be moderate, but for active sensors or systems, it will be high.

Specific Propulsion Systems Best Suited for Autonomous Vehicles   Most commercial transports are capable of carrying more than 250 passengers and a minimum crew of 12 with an extra fuel capacity close to 2 h. These commercial transports are powered with gas turbine–based turbofan engines and turbojet engines that can meet their thrust requirements. These turbofan engines come with different thrust ratings and offer high reliability, improved structural integrity, and ultrahigh efficiency in terms of fuel consumption per lb of gross weight including the passengers and crew, baggage, food and drinks, aviation fuel, and miscellaneous essential items. Practical examples of turbojet and turbofan engines widely deployed for unmanned autonomous vehicles are briefly mentioned as follows:

- British UVs deploy Rolls-Royce turbofan systems.
- British Aerospace UCAV Mantis, which weighs around 11,000 lb, uses turbojet engines.
- Russians have used the MIG-21 jet engines in their full-size UAVs and UCAVs. However, these engines suffer from reliability and maintenance problems according to published reports.
- Chinese UAVs and UCAVs deploy turbojet engines, which are strictly based on borrowed technology developed by Russians for their jet fighters.
- Turbojet engines (WP-13) are capable of providing thrust levels in excess of 9600 lb that are currently used from Soar Eagle and Global Hawk vehicles. Note that these two UAVs are widely used for undertaking high-altitude, long-endurance (HALE) surveillance and reconnaissance missions.

Research studies performed by the author seem to indicate that the sensor weight and estimated power consumption for the EO/EM systems can be summarized as follows:

Estimated Weight and Power Requirements for EO/EM Sensors aboard the Vehicle:

- X-band radar with target detection, acquisition, and target tracking (125 lb/65 kW)
- 10.6 μm laser illuminator including power supply (25 and 20 kW)
- Surveillance receiver (12 lb/20 W)
- Reconnaissance receiver (12 lb/20 W)
- Communication receiver (10 lb/25 W)
- Electronic warfare receiver (16 lb/26 W)
- IR missile warning receiver (15 lb/30 W)
- ECM or threat warning receiver (20 lb/30 W)
- Digital radio-frequency measurement (DRFM) equipment to counter radar threats (15 lb/22 W)
- CW noise jammer to jam enemy threat radars and RF missiles (30 lb/45 W)
- Pod-mounted IR deceptive countermeasure system (35 lb/120 W)
- Mini-RPV with 1 kW noise capability (18/150 W)

Note: The estimated weight and power consumption as mentioned above are strictly based on the assumption that state-of-the-art technology has been used for discrete components. Furthermore, IR and RF circuit components and other related elements should be used with minimum insertion loss, VSWR, and minimum weight, size, and power consumption. Finally, these estimates may have errors ranging from ±5% to ±10% approximately.

Summary

This chapter focuses on the capabilities of UAV technology and onboard sensor requirements to perform specific tasks. State-of-the-art vehicle design, high-speed computers for simulation activities, and most efficient and complex algorithms will be the essential elements and must be given serious consideration in the design of UAVs.

The latest tactical publications and report indicate that the Boeing X-47 B presents a classic example of unmanned combat autonomous vehicle. This aircraft took off from the carrier deck and demonstrated a glimpse of the autonomous concept during the tests conducted after the takeoff. This aircraft was controlled by the carrier deck operator using computer keyboard commands and mouse clicks. From the carrier deck, the human operator orders the aircraft to take off from the deck and perform specific tasks.

The X-47 naval aircraft flies at subsonic speeds and can provide an operating range in excess of 2000 nm. However, to demonstrate its capability, the aircraft needs to display the situational awareness for 100% autonomous capability. Encouraging signs of UAV capability are shown in U.S. automotive industry publications. These automobile industry publications reveal the use of EO sensors, computers, displays, smart materials, and dashboard instruments that will illustrate autonomous driving capability.

Studies performed by the author indicate that smart materials offer high stiffness at elevated temperatures exceeding 200°C, remarkable structural integrity under harsh atmospheric environments, and smooth control during atmospheric flights. Some rare earth materials are most useful in the development of micro- or minicomponents such as gyros, accelerometers, motion controllers, and fluidic actuators. The major benefits of smart materials include reductions in weight, size, and power consumption. Use of such materials for UAVs will offer many benefits including process guidance, vehicle stability control, smooth vehicle flight, and reliable flight performance of the vehicle. Combining the MEMS-based gyros with accelerometers can yield a backup means for the GPS. These gyros yield the highest accuracy in the instantaneous measurement unit (IMU). Furthermore, the MEMS accelerometers are widely used in multiple applications such as embedded systems, GPS sensors, precision IMU systems, systems with a full range of directional motions, and microelectronic sensors with interface compatibility.

Smart materials have been used in designing the fluidic actuators that can be embedded directly into the control surface of the vehicle. Fluidic actuators play a critical role in controlling the aerodynamic performance of the UAV. Essentially, these actuators improve lift force, decrease the drag force, improve the heat transfer efficiency, and smooth aerodynamic controls.

The role of electronic switch modules and programmable controls in conjunction with VFX displays is very critical because these elements assist the ground operator to get the most out of the systems deployed. The latest high-performance software offers flexible and scalable failure simulation. In computer simulation efforts, serious considerations must be given especially to vehicle maneuvering capability, vehicle reliability, vehicle overall performance, and collision avoidance capability. In other words, vehicle simulation efforts must focus on high-fidelity, multibody dynamics, monitoring of the critical operating data from the EO and EM sensors, and navigation capability in adverse atmospheric environments while the vehicle is flying at different speeds.

Evidence of supervisory control of a UAV can be demonstrated by using appropriate models. Efficient methods must be developed to monitor the activities of various system components, progress of the task selection, situational awareness, and decision-making planning actions. The L-PRISM method must be integrated with the multi-UAV control system known as TSM and system states to achieve their future goals. Furthermore, the role of three display components such as the state diagram, the timeline, and the payload viewer, as illustrated in Figure 6.4, must be examined carefully to achieve better understanding of autonomous vehicle performance. The typical payloads are the real images coming from the EO and REM sensors aboard the vehicle. Note that the L-PRISM technique allows the human operator sitting in the ground control station to monitor, inspect, evaluate, and manipulate the activities of multiple UAV aircraft in real time, situation assessment, tactical decision-making function, strategic planning, and immediate action required to control battlefield conflicts. EO and EM sensors aboard the autonomous vehicle such as SAR, FLIR, laser illuminator, multispectral camera, video camera, RF jammer, and RF receivers are described in great detail in this chapter with emphasis on reliability, collision avoidance tactic, low radar cross section (RCS) to avoid detection by enemy radars, and other tactical related sensors.

Critical design features of the UAV such as stealth aspect, locations of EM and EM sensors, high-resolution optical cameras, pod-mounted and pallet-mounted sensors, and installation location of propulsion systems are summarized with emphasis on reliability, vehicle safety, and security. Deployment of pod-mounted and pallet-mounted offensive and defensive weapons or sensors requires a trade-off study in terms of weight, size, and aerodynamics of the vehicle. Critical design parameters of the UAV are given serious considerations to the structural safety and security of the vehicle while operating over hostile territory to collect intelligence, reconnaissance, and surveillance data.

The UAV is equipped with laser-guided Hellfire missiles, RF and IR jamming equipment, passive and active electronic warfare systems, and other compact, lightweight weapons to assure the vehicle's safety.

References

1. NAVY's X-47 B UAV, Automatic landing and takeoff from the carrier deck, *Defense Electronic Technology Journal*, 17–18, March 2015.
2. A.R. Jha, *MEMS and Nanotechnology-Based Sensors and Devices for Communications, Medical and Aerospace Applications*, CRC Press, Taylor & Francis Group, New York, 2008, pp. 276–279.
3. A.R. Jha, *MEMS and Nanotechnology-Based Sensors and Devices for Communications, Medical, and Aerospace Applications*, CRC Press, Taylor & Francis Group, New York, 2008, pp. 117–119.
4. J. Harris, Autonomous changing everything, *Electron Product Company*, 58(2), 12–15, June 2015.

5. US Navy, NASA, and JPL Pasadena (CA), Real time, high fidelity simulation of autonomous vehicle dynamics, *Aerospace and Defense Technology*, 36–37, May 2014.
6. M. Patzek et al., Supervisory control state diagrams to depict autonomous activity, *Aerospace and Defense Technology*, 30–31, May 2014.
7. NASA editor for Tech Brief, Queuing model for supervisory control of unmanned autonomous vehicles, *Aerospace and Defense Technology*, 34–37, May 2014.
8. J. DiVita et al., *A Queuing Model for Supervisory Control of Unmanned Autonomous Vehicles*, Space and Naval Warfare Systems Center Pacific, San Diego, CA, September 2013.

7

SURVIVABILITY OF UNMANNED AUTONOMOUS VEHICLES

Introduction

This chapter strictly focuses on the cutting-edge technology, sensor design concepts, and fast as well as efficient software functions that significantly enhance the survivability, safety, and security of unmanned autonomous vehicles (UAVs) and their contents. Since the UAV or platform is a complex and costly vehicle, it should be operated and controlled by high-speed computers and sophisticated electro-optical (EO) and electromagnetic (EM) sensors under the guidance of an experienced ground control operator. This ground control operator must be a seasoned fighter/bomber pilot with multiyear combat-related experience.

Critical Issues and Factors Responsible for UAV Survival

The author will discuss the critical technical issues and factors that are responsible for the survival, safety, and security of the UAV and its contents. Survivability studies performed by the author reveal that stealth features are critical requirements that strictly depend on the fuselage structural factors, control surfaces, radio-frequency (RF) radar cross section (RCS), infrared (IR) signature, jet engine exhaust parameters, and the location and size of the pod-mounted EO and EM sensors and missiles. Other factors such as radar absorption paint, deployment of smart materials, and direction of hot exhaust gases exiting from the nozzle also play critical roles in the survivability of the UAV.

Stealthy Fuselage Features and Control Surfaces

Stealth features of the UAV and RCS must be given serious consideration to avoid detection by the enemy radar. These two requirements will significantly reduce the RF signature, which is essential for the survivability of the UAV. Experienced battlefield commanders believe that stealthy features are very essential for the deep penetration of fighter/bomber platforms in heavily defended enemy air corridors. Various techniques will be discussed to achieve significant reduction of RF and IR signatures to avoid detection and tracking by enemy EO and EM sensors.

Assuming that these UAVs are required to penetrate deep into the heavily defended enemy territory, they must operate like hunter–killer UAVs. There should be a balance between the number of weapons carried aboard the vehicle and the endurance capability and sensor payload including the pod-mounted compact weapons. It is extremely important to point out that any imbalance will impact not only the UAV endurance capability but also the ability of the vehicle to reach the center of the battlefield in the allocated time frame. According to military and battlefield experts, the Air Force's MQ-9 Reaper will be a perfect match for the two driving forces mentioned in the hunter–killer concept [1]. This particular concept is necessary to provide new precision strike capability on the global war on terror. It will be highly desirable for hunter–killer platforms to undertake precision strike capability using advanced EO and EM sensors as well as covert intelligence collection, surveillance, and reconnaissance (ISR) functions.

It will be interesting to point out that military experts can classify the U.S. Army's Warrior UAV as a hunter–killer platform, because it has been already optimized for deep penetration capability in heavily defended enemy territory and for undertaking a combination of attack and ISR missions. Note that precision attack missions in the hunter–killer platforms must be equipped with the following sensors with high accuracy:

- EO and IR sensors
- Hellfire IR laser-guided missiles
- Side-looking radar (SLR) capable of seeing through rough clouds, smoke, and airborne obstacles

- Covert communication relay
- Miniaturized weapons such as Raytheon's Pyros laser-guided smart weapon
- Uncooled forward-looking infrared (FLIR) system with narrow field-of-view capability

Note that a UAV equipped with stealthy features, advanced EO and EM sensors, and hunter–killer capability will provide both the high-kill probability and high probability of survivability of the UAV or platform. Such UAVs with stealthy features will play a critical role in future wars without pilots in cockpits.

Deployment of Smart Materials for Stealth Capability: Comprehensive research studies undertaken by the author seem to reveal that smart materials possess unique properties that will provide not only high mechanical integrity but also significant reduction in the weight and size of the components.

Smart Metals and Alloys: Studies performed by the author indicate that smart materials come in various categories, namely, MuMETAL®-based meals, composite alloys, composite plastic materials, and rare earth–based metals and alloys. These smart materials have unique aerospace, aeronautical, and mechanical properties best suited for UAV components, because such smart materials play an important role in significant reduction in weight and size and provide major improvement in platform reliability and survivability. Our research studies further indicate that IR-guided missiles and laser illuminators require specific IR window and dome materials that will minimize false targets generated by solar reflections from the clouds and terrestrial objects while optimizing the sensitivity over the desired IR spectral window.

Smart Optical Materials   Smart optical materials are used for the design of IR windows and domes. Smart optical materials for the windows must be selected for laser-guided bomb or IR missile receiver dome. Furthermore, optical materials for the IR dome or window must experience minimum cloud reflections and atmospheric absorption and deflections to maintain high pointing accuracy. Advanced IR window materials include diamond, $ZnSe$, and Al_2O_3. The last material is not suitable for high-power laser sources because of poor thermal conductivity. Our studies reveal that the diamond window offers low insertion loss, highest thermal conductivity, high dissipation capability, and improved resistance against thermal shock. Research studies further indicate that deposition of surface compressive layers on the front will lead to significant increase in strength and ductility, which will avoid catastrophic failure of the window. Comprehensive examination of smart window materials indicates that the threshold of the velocity for observable impact damage in the window material is directly proportional to the damage parameter D, which is the product of the fracture toughness and the elastic wave velocity.

Stealth Technology Vital for UAV Survival

For deep penetration of the UAV, the UAV platform must not be detected by enemy sensors. The deployment of stealth technology is extremely essential for the survivability, protection, and safety of the most expensive, heavily armed, and fully equipped UAV. Furthermore, both RF and IR signatures must be kept as low as possible to avoid detection by radar and IR receiver. The reduction techniques for RF and IR signatures will be discussed under appropriate sections. The vehicle RCS must be small enough to avoid detection and tracking by enemy radar. This will require that fuselage and control surface dimensions be as small as possible; the fuselage must avoid sharp edges, pod-mounted sensors, or missiles; and the entire vehicle must be painted black using highly absorbent iron-ferrite paint of appropriate thickness.

RF Signature Reduction Techniques: Preliminary survivability studies undertaken by the author show that the radar signature should be well below –40 dB/m² to avoid detection by high-power enemy radar systems. The latest RCS reduction techniques are commercially available and will be discussed briefly.

RCS Reduction Techniques by Vehicle Structural Design Concepts

Stealth fighters and bombers designed and developed from 1960 to 1990 have demonstrated significant reduction in IR signatures and the RCS strictly due to unique concepts for the fuselage, control surfaces, jet engine exhaust configuration, undersized vertical stabilizers, and other aerodynamic features such as shadowy shaped air inlets [2]. Aerospace scientists believe that stealth technology touches most every major aerospace company. In addition, much-published military airplanes now in service are showcases of earlier generations of stealth technology. The SR-71 reconnaissance aircraft and the B-1B bomber seem to incorporate stealth features, according to major aerospace research and development companies.

Published literature [2] reveals that previous stealth-based air-crafts are constructed of radar-transparent and radar-absorbing materials (RAMs). In addition, engines, missiles, and other highly radar-reflective metallic components were located at appropriate locations in the fuselage section. The survival of such aircraft requires the onboard electronic and RF equipment to minimize their EM emissions to avoid detection by sensitive enemy receivers. Advanced computer-based flight control systems ensure that these inherently aerodynamically unstable moving platform designs can fly safely. Special flight control surface mechanisms make up for the under-signed vertical stabilizers.

RAM and Radar-Absorbing Paint for RCS Reduction: RF engineers and anechoic chamber designers believe that one can absorb 90% of the RF energy using appropriate RAM thickness and when installed at suitable locations on the fuselage surface. For best results, RAM is installed on flight control surfaces, auxiliary air duct, deep section of RAM on leading and trailing edges, inlet ducts lined with RAM, and other exposed sections of the airframe.

Iron-ferrite paint of appropriate thickness when applied to the aforementioned surfaces will absorb RF energy ranging from 35% to 45%. Radar-absorbing paint (RAP) should be applied to essential surfaces of the UAV fuselage such as control surfaces, inlet air ducts, vertical stabilizer, trailing and leading edges, and other critical surfaces. Note that the absorption capability of the RF paint is strictly dependent on the paint composition, thickness of the paint, and RF absorption capability in terms of db/mil thickness of the paint for a specified RF spectral window. Note that special paints are available for optimum effectiveness over appropriate RF spectral ranges such as 4–6, 6–8, or 8–12 GHz. Most threat radars operate in X-band frequencies ranging from 8.2 to 12.4 GHz. Some Soviet threat radars operate between X-band and K_u-band frequencies, which makes RAM effective near borderline. In such situations, the paint composition must be modified, and samples with flat surfaces should be painted and evaluated in an anechoic chamber for the observation of the paint effectiveness. Preliminary studies undertaken on the effectiveness of radar-absorbing energy paint seem to indicate that a paint thickness of 0.010–0.015 in is adequate to absorb most of the RF radiation coming out from the RF source operating over 8.2–12.4 GHz. However, the same RAP might indicate higher radiation-absorbing capability over the 12–16 GHz frequency spectrum. Stealth technology is classified as low observable (LO) technology and considered a passive electronic countermeasure tactic. An airplane with low scattering surface and using RAP has demonstrated an absorbing capability exceeding 90%. Such an aircraft is considered invisible. Use of fiberglass for aircraft surfaces and then painted with special RAP will ultimately have the RF absorbing capability better than 95%.

Nanopaint Effectiveness for UAV Platforms: Aerospace scientists feel that the latest "nanostructured" paint coating for stealth aircraft cannot be seen in day and night and also that they are undetectable by threat radar during the day. Paint coating made of carbon nanotubes (CNTs) can be used to cloak an object in dark environments, making it indistinguishable from the night sky. Research studies performed by other scientists confirm that paint consisting of black CNT elements demonstrated unique properties, including significantly improved mechanical strength, higher electrical conductivity, and improvement in paint quality. The CNT-based RAP has a diameter of a few nm (10^{-9} m) and is capable of absorbing a spectrum of light or RF spectrum from radar waves through visible light spectrum through UV spectrum. According to CNT paint scientists, the presence of vertical CNTs on the surface of a 3D object makes the objects appear completely flat and black. Because these objects are completely black, they are known as the blackest material. The RAM coating is made of CNTs, which are long straws of pure carbon. Note that each unit has a diameter of a few nanometers and is capable of absorbing a broad spectrum of RF radiation. It is interesting to point out that the nanotube-coated agents neither reflect nor scatter light or RF signal. Under such conditions, if the aircraft is coated with nanopaint and illuminated by a microwave radar beam, nothing will bounce back and it would appear as if nothing was there.

Simple and less expensive RAP is also available and is known as "iron ball paint." This particular paint contains tiny spheres and is known as nanosphere paint. RAM manufacturers do not provide the paint composition details because they consider this to be a trade secret. There also exists a stealth paint that absorbs the IR radiations from a laser source or from an IR source.

Genetic Algorithms to Design the RAM: Research studies undertaken by the author seem to indicate that the use of genetic algorithms should be explored for the design and development of RAMs. These algorithms would identify the composite method to design the RAMs with minimum cost and complexity. If the military authorities move the composite materials from ship surfaces and superstructures, then the frequency-selective surfaces and circuit analog absorbers should be embedded into the composite material. For full details on this technology, RAMs should be studied and evaluated in great detail to counter frequency radars, which changes radar frequency instantly to avoid the presence of radar. This technique is very attractive in the assessment of RAM paint effectiveness.

Techniques Currently Available for RCS Reduction [2]

Comprehensive research studies conducted by the author reveal that four potential methods for RCS reduction are currently available, which can be summarized as follows:

1. Shaping active loaders
2. Passive loading concept
3. Distributed loading technique
4. Comprehensive shaping of fuselage structure

As mentioned earlier, the RCS of the vehicle is strictly dependent on the target size, structural shape, and materials used for the airframe. Research studies further reveal that passive and active loading techniques realize a significant reduction in RCS, but at the expense of higher machining and labor costs. However, the use of genetic algorithms could lead to optimum loading of RAM layers in minimum duration. Regardless of how one does the loading of the structural elements, labor and machining costs can be controlled. Note that loading requires more time and labor, which will undoubtedly increase vehicle costs. However, the program managers who are deeply involved in tracking the progress and in the management of stealth features feel that shaping and loading are most promising in reducing the backscattering signal of the target. But it can redirect the radiation through specular reflections, thereby increasing the probability of detection through bistatic loading. It should be remembered that active and passive loading techniques aim to reduce the scattering signals from the hot spot regions.

Note the third loading method ("distributed loading technique") covers the aircraft surface with a RAM that has imaginary components of permittivity (e_r) and permeability (μ) associated with the electric or magnetic fields of the radiation coupled with the metallic material properties. To optimize the effectiveness of this particular method, more emphasis should be placed on genetic algorithms to achieve circuit analog and frequency-selective surfaces.

Note that inexpensive RAMs are available in the market which are known as conductive polymers or as special composite materials. In the case of RAMs, the resonance frequency of the absorbing material is tunable through the variation of the resistance and capacitive elements of the absorber. Graduate and undergraduate students should understand that intrinsic impedance (Z_0), which is known as characteristic impedance, is the ratio of the electric field vector to magnetic field vector and can be written as

$$Z_0 = [\mu/e_r]^{0.5} = [376.6] \text{ or } [377] \, \Omega \qquad (7.1)$$

Any RAM with $Z_0 = 377 \, \Omega$ will not reflect microwave energy but will absorb most of the radiation impinging on it. This is the basic characteristic of the intrinsic impedance of the RAM. For optimum absorption capability, the RAM thickness must be $\lambda/4$. For example, for a RAM to absorb most of the radiation over a spectral bandwidth of 10–12 or 10–15 GHz, its thickness must be close to 0.300 in. This RAM thickness is based on the lowest frequency of the spectral band.

Latest Paints Best Suited for RCS Reduction

The following are paints best suited for RCS reduction:

- Dodge vapor paint
- Iron ball paint absorber (contains nanoparticles)
- Radar-evading camouflage paint (contains microscopically minute particles invented by Germans)
- Black RAP (widely used for fighter/bombers)
- New stealth nanopaint
- Iron-ferrite paint

IR Signature Estimation and Reduction Techniques

IR signature reduction must be given serious consideration when an aircraft is required to undertake deep penetration covert tactical missions. IR emissions must be adequately suppressed to avoid "lock-on" by the enemy IR missiles. The hot exhaust gases from the UAV jet contribute the most IR signature, which should be suppressed for the survivability and safety of the aircraft and its contents. Various techniques to reduce IR signatures from the vehicle will be discussed for the protection and survivability of the UAV. It is interesting to point out that most high-performance UAVs are expected to use jet engines with output thrust ranging from 11,000 to 15,000 lb [3].

Note that the overall IR radiation level and lock-on range can be determined by the General Electric (GE) Scorpion software program. Essentially, this program uses molecular band models to describe the spectral distribution of the average intensity and the widths of the spectral lines within the spectral bands. These models are based on quantum theory and this software program takes into account all relevant geometrical parameters that define the exhaust system exit, exhaust gas exit location, and orientation of the hot plume from the jet engine.

Furthermore, the UAV airframe comprises various structural elements such as nose section, fuselage, wings, control surfaces, and other elements such as the elevation stabilizer. These elements will have different surface conditions, temperatures, and materials. Even these elements contribute to the overall IR signature. Note that the emissivity of the material plays a key role in the IR signature

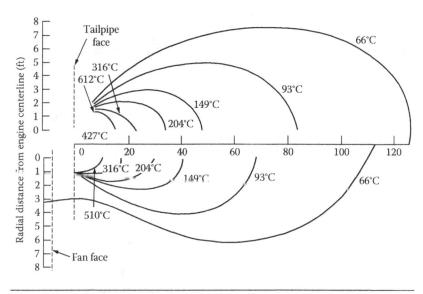

Figure 7.1 Contours of exhaust temperature for jet engine under cruise environments.

prediction, and it is a function of surface conditions, temperature, and IR emission wavelength. For a given metal surface temperature, the emissivity (e) is slightly affected by the aspect angle and its value could vary from 0.9 at 0° aspect angle to 0.7 at 90° aspect angle. Under these variable operating conditions, the accurate estimation of the IR signature from the jet engine tail pipe could be extremely difficult, if not impossible.

Preliminary results obtained from this program seem to indicate that the radiation intensity (WSr) during afterburner mode from a jet engine tail pipe is roughly 25 times that of an exhaust plume under normal cruise flight conditions. Comprehensive studies undertaken by the author indicate that fundamental expressions need to be derived for better understanding of IR radiation intensity (Figure 7.1).

Thermal Expressions Used in the Calculation of IR Signature

Research studies conducted reveal that the total radiation level from the jet engine tail pipe contributes to the IR signature. Other IR signature contributions from various sources or aircraft elements are relatively low as the jet exhaust temperature increases. The expression for the IR radiation level from the jet engine tail pipe can be written as follows:

$$R = [e\ 0\ T^4] \tag{7.2}$$

where

    e is the emissivity of the metal, which is a function of temperature and wavelength

    0 is the constant (5.67×10^{-12})

    T is the exhaust gas temperature or jet tail temperature (K)

The expression of the radiant emittance in watts (W) can be written as follows:

$$W = [e\ 0\ T^4]\ W/cm^2 \tag{7.3}$$

Sample Calculation   Assuming an emissivity of 0.9 and exhaust gas temperature of 900 K and inserting the given parameters in Equation 7.3, one gets the magnitude of W as

$$W = [0.9 \times 5.67 \times 10^{-12} \times 900^4] = [3.343]$$

$$= [3.343]\ W/cm^2$$

Assuming the exhaust pipe diameter is 12 in or 30.34 cm, one gets the magnitude as

$$W = [3.343 \times 22/7 \times 30.48^2]\ W$$

$$= [2444]\ W \text{ at the tail pipe exit} \tag{7.4}$$

These calculated values are valid under static conditions. Note that these values will change under dynamic conditions, which means that the aircraft is moving under the influence of roll, pitch, and yaw angles.

IR Radiation Intensity (IR Signatures) at Various Elements of the UAV

Under dynamic conditions due to aircraft motion and maneuvering, the IR signature will experience changes as a function of roll, pitch, and yaw angles. It should be remembered that at constant altitude and aircraft speed, one can expect smooth flight conditions. Under such conditions, which are described as cruise flight conditions, there

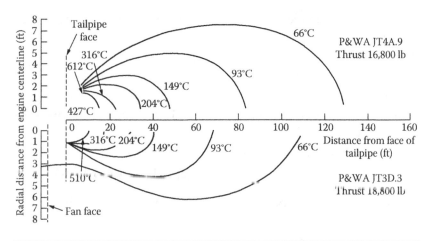

Figure 7.2 Contours of the hot exhaust gases from the aircraft tail of the commercial transports.

will be constant IR radiation intensity. Figure 7.2 shows contours of exhaust temperatures for turbojet (18,000 lb thrust) and turbofan engines under cruise conditions. The simulated IR signature or radiation intensity (W/Sr) from a commercial jet engine with and without atmospheric attention at various aspect angles can be seen in Figure 7.3. For these simulation results, the jet engine thrust is assumed to be around 15,000 lb. Notice that the radiation intensity level or IR signature is maximum at the tail location, because the engine hot exhaust exits at that location. However, the IR radiation level or IR signature at the aircraft nose is significantly less than 2%.

The IR radiation intensity simulation data obtained under aircraft roll and pitch conditions with and without atmospheric attenuation are much less than 16% under pitch condition and under atmospheric attenuation environments. However, at the aircraft tail location, the IR signature is approximately less than 42%, as illustrated in Figure 7.4. Note that the IR signature under roll conditions and with atmospheric attenuation is close to 13% at the tail's aspect angle. The IR radiation intensity or the IR signature undergoes significant changes at all aspect angles and under pitch conditions. Even though no simulation data were taken, the author still predicts that the IR signature will remain fairly constant at small yaw angles, because the direction of hot exhaust gases exiting from the aircraft tail nozzle will be fairly constant, regardless of the aspect level. These are the most important observations one can make from the curves shown in Figure 7.4.

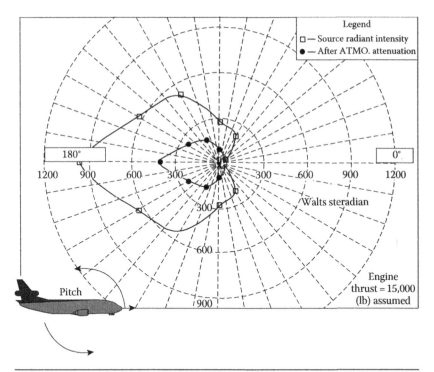

Figure 7.3   Simulated IR signature or radiation intensity (W/Sr) from a commercial jet engine with and without attenuation at various aspects angles.

All these figures indicate that the maximum RF signature occurs at the tail of the vehicle, while the low-intensity IR signature occurs in the forward flight sectors [3]. The peak IR radiation intensity of 1080 W/Sr occurs at the tail aspect angle of the aircraft irrespective of the roll and yaw angles. Note that the IR signatures undergo radical changes as a function of pitch angle. It is interesting to point out that the IR radiation intensity at the tail aspect angle will be significantly reduced if the downward pitch angle is greater than 15°–25°. Under these pitch angles, it will be extremely difficult for the IR missile to lock on the UAV platform.

Comprehensive studies performed by the author on the IR signature of the exhaust tail pipe reveal that moderate reduction in the IR radiation intensity at the tail pipe begins to increase when the pitch angle varies from 0° to 20° at altitudes exceeding 5000 ft. The studies further reveal that the maximum IR signature occurs over the 2.8–3.2 μm spectral bandwidth, when the tail pipe temperature ranges from

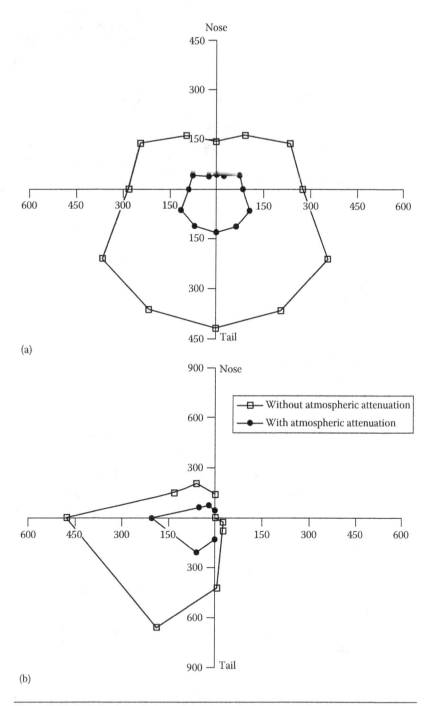

Figure 7.4 Radiation intensity from a jet engine under (a) aircraft roll and (b) pitch conditions.

900 to 1065 K. Computer simulation data indicate that the maximum IR signature occurs when the jet engine is operating in the afterburner mode. Under these conditions, the IR radiation is confined to the 1.3–1.8 μm spectral range. Finally, the emissivity of the exhaust pipe is a complex function of wavelength and temperature, which can be written as $\lambda = f(\lambda, T)$, where λ is the wavelength in μm and T is the jet engine exhaust's absolute temperature in Kelvin.

IR Signature due to Aircraft Skin Temperature [3]

When an aircraft is moving at high speeds, its fuselage, wing surfaces, control surfaces, and pod-mounted sensors generate skin temperatures that contribute to overall IR signatures. However, these skin temperatures are significantly lower than tail pipe exhaust temperatures.

The expression for the skin temperature can be written as

$$T_{skin} = [216.7 (1 + 0.16M)]K \tag{7.5}$$

where

T_{skin} is the skin temperature (K)
M is the speed of the aircraft in Mach (666 mph)

Computed values of skin temperature (T) and peak radiation wavelength (λ_{peak}) as a function aircraft speed and peak wavelength are shown in Table 7.1.

IR Energy Generated by Various Aircraft Elements

Any IR energy generated from the aircraft skin and control surfaces will be added to the IR energy from the jet engine plume's IR energy

Table 7.1 Skin Temperature and Peak Radiation Wavelength as a Function of Aircraft Speed

UAV SPEED, M (MPH)	T_{SKIN} (K)	PEAK WAVELENGTH, λ_{PEAK} (μM)	FLIGHT STATUS
0.5 (330)	234	12.39	Cruise
0.75 (499)	243	1193	Cruise
1.00 (666)	251	11.55	Cruise
1.25 (832)	260	11.1	Afterburner
1.5 (999)	269	10.78	Afterburner

over certain spectral bandwidths, leading to overall IR signature of the UAV. In addition, IR energy generated from pod-mounted sensors and weapons will contribute to the overall IR signature of the vehicle over the specified spectral bandwidths as shown in Table 7.2. More than 90% of the IR signatures are contributed by jet engine exhaust gases and the rest are contributed by fuselage, wings, control surfaces, and pod-mounted sensors and weapons over specified spectral bandwidths, as shown in Table 7.2.

Preliminary thermal calculations show that a tail pipe temperature of 900 K transmits only 16.50% of the total IR radiation in the 3.00–2.00 μm spectral range [3]. On the other hand, a tail pipe temperature of 1000 K delivers a radiant emittance of 20.4% over the spectral bandwidth of 3.00–2.00 μm. However, at the same exhaust temperature of 1000 K, the IR radiant energy is significantly reduced to 11.2% over the slightly wide IR spectral bandwidth of 3.00–2.500 μm. This means that the wider the spectral bandwidth, the lower the percentage of the IR energy or IR signature and vice versa.

In case the UAV is operating in the afterburner mode, most of the RF energy will be confined in narrow spectral bandwidths as shown in Table 7.3.

These calculations confirm the fact that a maximum percentage (20.6%) of the IR energy or IR signature is confined to a narrow IR

Table 7.2 Percentage of Radiant Energy or IR Signature over the Specified Spectral Bandwidths

TEMPERATURE (K)	SPECTRAL BANDWIDTH, $\lambda_1 - \lambda_2$ (μM)	PERCENTAGE OF TOTAL RADIANT ENERGY	REMARK
500	6 – 5	11.30	
900	3.20 – 2.00	16.30	
1000	3.00 – 2.00	20.40	Max. IR signature
1000	3.00 – 2.50	11.20	

Table 7.3 IR Radiation Emittance over Specified Spectral Bandwidths in Afterburner Mode

TEMPERATURE (K)	SPECTRAL BANDWIDTH, $\lambda_2 - \lambda_1$ (μM)	PERCENTAGE OF TOTAL RADIANCE EMITTANCE
2000	1.50 – 1.00	20.6
2000	1.50 – 1.20	13.3
2000	1.50 – 1.25	11.2

spectral range of 1.50–1.00 μm similar to an observation noticed under cruise conditions.

MAM Technology for Small and Lightweight Munitions

Since the UAV is a complex and very expensive vehicle, the components, devices, and tactical munitions for these vehicles must be carefully designed with particular emphasis on the weight and size of the components and tactical munitions. This design condition requires deployment of multiple additive manufacturing (MAM) technology [4] in designing the tactical munitions, pod-mounted systems, and housings for externally mounted EM and EO sensors. Comprehensive IR and RF signature studies undertaken by the author indicate that deployment of MAM technology could result in reduction of IR and RF signatures. Use of MAM technology in the design of jet engine nozzles could reduce the IR signature from the hot jet engine exhaust gases. Furthermore, use of MAM technology in the design and development of tactical munitions and pod-mounted sensors can significantly reduce the RCS signature, thereby indirectly providing the survivability and physical security of the UAV. Various advantages of MAM technology can be briefly summarized as follows:

- Weight and size reduction
- IR signature reduction
- RF signature reduction
- Avoidance of enemy radar detection and tracking
- Avoidance of IR missile lock-on

All these benefits of MAM most likely would contribute to the survivability, security, and safety of the UAV when undertaking dangerous combat missions deep in enemy territory.

Specific Details on MAM Technology

Preliminary studies of the current manufacturing technologies seem to indicate that MAM technology offers potential benefits in the design and development of missiles, pod-mounted EO and EM sensors, small tactical munitions, and other conventional weapons best suited for UAVs [4]. Battlefield commanders and defense consultants believe that this particular technology is most attractive where stealth features, survivability, safety, and security are of prime consideration. This technology offers use of various metallic materials that are highly resistant to chemicals and extreme thermal environments. Essentially, MAM technology plays a critical role in fused deposition modeling (FDM) and selective laser sintering (SLS). Both the FDM and SLS processes could reduce material and labor costs in the manufacturing of pod-mounted sensors and tactical munitions. Also, both the FDM and SLS techniques provide high-quality materials that are highly resistant to chemical and extreme thermal environments.

Highlights of FDM Technology: FDM technology can play a significant role in the manufacturing of small tactical munitions with minimum cost and complexity. These munitions are best suited for conventional aircrafts, unmanned aerial vehicles, and unmanned combat aerial vehicles (UCAV).

Additive manufacturing technology (AMT) deploys the latest 3D printing technology that offers high-precision components and most compact parts with high reliability and tight tolerances. Such parts and components are ideal for UAVs and also for conventional fighter/bomber aircraft.

Raytheon Company gives serious consideration to MAM technology. Their weapon designers have selected MAM technology for manufacturing of small tactical weapons called "Pyros," which is considered ideal for unmanned cruise aerial vehicles. The latest and the newest Pyros laser-guided missile shown in Figure 7.5 is lightweight, compact, and precise and is considered the most effective weapon.

3D printing technology allows direct prototyping to actually using additive manufacturing parts on the production component or a critical system part. Following such a unique manufacturing technique not only reduces labor costs but also the production time of critical components. In other words, this technique saves labor costs as well as production time, thereby significantly reducing the manufacturing cost of the system.

Comprehensive evaluation of this technology will confirm that deployment of MAM technology during the Pyros prototyping phase and into the final production of the part including the FDM and SLS procedures will take minimum time and cost.

Figure 7.5   Installation of Pyros under the UAV platform.

Furthermore, both FDM and SLS procedures would yield high-quality parts, which are best suited for the manufacturing of UAV and UCAV platforms. Furthermore, the UAV components produced using this technology are highly resistant to adverse chemical effects and extreme thermal environments.

The FDM process works via a heated nozzle that extrudes a material layer by layer while the SLC process works via a bed of powdered nylon and a carbon dioxide laser operating at 10.6 µm wavelength which sinters a material layer by layer. These two processes grow parts from the ground up, which eliminates other manufacturing complexities such as machining and emulation. These procedures yield precision tactical weapons with homogenous structural parts of minimum weight and size.

Current machine shop operators believe that the drilling and milling of a "computer numerical control" (CNC) machine is extremely costly and time consuming. In addition, for complex munition components or parts, it is relatively cheaper and quicker for small tactical munitions such as Pyros or even for standard missiles to use additive manufacturing for munition components that require tight machining tolerances. According to suppliers, manufacturing experts, and Raytheon engineers, extreme hard work is required to design and develop Pyros to meet the feasibility of assembling the munitions at the earliest time and within the fixed budget contract.

Raytheon program managers believe that AMT plays a key role in integrating the stated design features directly into the proposed geometry, including the attachment features, fitting mounting brackets, and control surface with minimum RCS requirements. A Raytheon program manager stated that the design and development are considered the most difficult tasks to achieve in one simultaneous and complete manufacturing cycle. In other words, the whole weapon system assembly can be completed with additive manufacturing and machining technology with tight tolerances. It is interesting to point out that AMT has allowed the Raytheon engineering team to consolidate multiple assembly features into one specific part and gave the engineering team full control over minute changes in control surfaces and tolerances.

Location of the Control Fins: Note that this particular weapon is a laser-guided missile, and attachment of control fins at the right place on the body of the missile is absolutely necessary. Since the control fins are critical for guiding Pyros, laboratory experimenting with their control surfaces should be on the forefront of design and development efforts.

Published articles indicate that AMT will play a critical role in weight and package size reduction. The article further indicates that Pyros is most suited for UAVs and UCAVs involved in armed attack and in deep penetration surveillance missions. These missions have typical payload requirements ranging from 5 to 75 lb. Raytheon program manager estimates that most of Pyros' weight comes from its warhead and therefore the weight must be subtracted elsewhere. Since material compositions of nylon are used in conjunction with the SLS, yield parts are light but require high mechanical strength and are highly resistant to harsh operating environments while incorporating more features than machining could feasibly achieve in a single manufacturing cycle [4].

3D Printing Technology

3D printing technologies have been recommended by computer experts and space scientists to reduce the weight and size of a complex system. As a matter of fact, 3D printing technology was used in the product version of a space flight hardware in January 2014.

SpaceX first flew its Falcon 9 with a 3D printed main oxidizer (MOV) body in one of the nine MERLI 10 engines. Note that the valve is used to control the flow of cryogenic liquid oxygen to the engine in a high-pressure, low-temperature, high-vibration physical environment. Other 3D printed spacecraft assemblies have been ground tested, but not yet flown in space, including high-temperature, high-pressure rocket engine combustion chambers and the entire mechanical space frame and propellant tanks for a small satellite weighing roughly a few hundred kilograms. As mentioned earlier, MAM technology can be used for several small tactical weapons that are most ideal for UCAVs and UCAV platforms.

MAM technology is also known as 3D printing technology, which deals with objects of any physical shape or geometry and objects that are produced from a 3D model. Furthermore, the 3D printable models can be created with a computer-aided design (CAD) package or via a 3D scanner or a plain digital camera using photometry software. Jet engine designers believe that 3D printing technology is best suited particularly in jet engine nozzles using smart materials to achieve high nozzle efficiency, improved mechanical integrity, and high nozzle throat efficiency under high-temperature and high-pressure environments.

Estimated Performance Parameters and Physical Dimensions of Pyros: Wikipedia articles describe the performance and physical data of small tactical munitions (Pyros) developed by Raytheon, which can be summarized as follows:

- *Estimated weight*: 13 lb
- *Maximum length*: 22 in
- *Best suited platforms for launch*: UAVs or Predator and Reaper aircraft
- *Brief description of the system*: Dual-mode, semiactive laser seeker that includes the global positioning system (GPS) and INS sensors as guidance systems

Potential Applications of Pyros Munitions   The following are potential applications of Pyros munitions:

- RQ-7 Shadow UCAV.
- *Old warhead weight aboard this vehicle*: 7 lb.

- *New warhead weight for the same vehicle*: 5 lb.
- New warhead was developed using AMT.

Potential Benefits of AMT The following are potential benefits of AMT:

- Significantly improved blast–fragment capability.
- Offers low collateral damage.
- *Drop time from 10,000 ft altitude to ground*: 35–40 s.
- AMT provides precision machining with extremely tight tolerance. Note that engine nozzles designed and developed using smart materials and AMT demonstrated high nozzle throat efficiency, improved reliability, shorter time to develop afterburner capability, and high mechanical integrity under extremely high-temperature and high-pressure operating environments.

Future Space System Deploying 3D Printing Technology: The new United Launch Alliance Vulcan launch vehicle, which is not expected to launch earlier than 2019 for evaluation of 3D printing technology. This program involves more than 150 parts, which include 100 polymer parts and 50 other parts. The AMT technology was demonstrated for particular application to rocket engines and spacecraft structure.

3D Printing Technology Enhances Pyros and GPS Performance: Comprehensive examination of 3D printed technology reveals that this technology significantly improved the performance of the GPS. With 3D coordinates for the GPS integrated with a laser-guided sensor, Pyros knows precisely where it wants to be. Furthermore, for moving targets or targets within the building, the Pyros is equipped with a semiactive laser guidance system with demonstrated accuracy of 1 m or 3.2 ft. A laser designator's optical energy reflected off the target is used by the seeker to guide the Pyros to the target. It is interesting to point out that all the target direction information is processed simultaneously, which makes it accurate for the fin movements in accordance with the GPS sensor and laser information which is considered most essential to the success of the UAV mission.

It would be highly desirable for Pyros design engineers to work closely with different component companies for possible iterations of Pyros munitions using the prototype and production capabilities of additive manufacturing. The engineering design team looks forward to reworking the guiding fins using AMT.

3D Printing Technology for Commercial Passenger Transports: In May 2015, Airbus management announced that its new Airbus A350 XWB will include more than 1000 parts to be manufactured using 3D printing technology. Until recently models were built by hand, which took more time and labor. Thereafter, the architects were forced to show the part drawings to their clients. But the clients wanted to see the product or part from all possible viewpoints of the design in space to see a clear picture of the designed part to make a firm decision in order to get the scaled models to clients in a small amount of time. It is very important to note that now the architects and architecture firms tend to rely on the 3D printing technology concept. Most likely this will be the future trend for the special part procurement cycle from now on. Engineering firms believe that 3D printing technology can reduce lead time of production cycle by 50%–80%, producing scale models up to 60% lighter than the machined models while being sturdy. It now appears that designs and scale models are only limited by a person's imagination.

3D printing experts believe that their technology offers significant improvements in accuracy, part production speed, saving of material, manufacturing tolerances, and quality control. Advocates feel that 3D printing technology has opened new doors to move 3D printing beyond the modeling process and ultimately into its manufacturing strategy. Note that contour crafting is the most useful feature of 3D printing technology, which can play a critical role in additive manufacturing processing technology. It should be remembered that contour crafting uses a computer control system. 3D printing technology permits layers of different materials. 3D printing technology is vital for rapid prototyping.

Examples of Success for 3D Printing Technology: Comprehensive research studies undertaken by the author on 3D printing indicate that this particular technology has been deployed in specific commercial, industrial, and research applications. Applications of 3D printing can be briefly summarized as follows:

- Production of customized food such as pizza, pastries, and candies using hydrocolloid (a substance that yields a gel when mixed with water at room temperature) in an AMT process has been accomplished.
- Research scientists have demonstrated the development of complex geometries in the shortest timeframe with minimum cost and complexity.
- 3D printing technology has been deployed in the world of clothing and shoes involving sophisticated designs of ladies' shoes and multicolor fashionable dresses.
- Manufacturing experts believe that 3D printing technology has a great future in the automobile, firearm, medical, sports, computer, and robot industries with minimum initial investment.
- 3D printing technology is best suited for manufacturing of steam turbine blades and turbojet engine blades. This is the only technology that will allow different alloys at the root and tip of the blade.
- 3D printing technology must be used where precision, accuracy, quality control, repeatability, and reliability are the principal design requirements.

Role of AMT in Making Jet Parts: GE is making a radical departure from the way it has traditionally manufactured aviation parts. The company's aviation division, the world's largest supplier of jet engines and turbojet engines, has decided to produce fuel nozzles by printing the parts with lasers rather than casting and welding the metals. The latest new technology known as additive manufacturing, which builds the part by adding ultrathin layers of material one by one, could transform how GE designers manufacture many complex parts that go into everything ranging from gas turbines to jet engines to ultrasonic machines.

Additive Manufacturing Technology Concept: AMT is considered the industrial version of the 3D printing technology that has already been deployed in the development of medical implants and in the production of plastic prototypes for engineers and product designers.

However, the decision to mass produce critical metal alloy parts for deployment in thousands of jet engines is a significant milestone for this technology. 3D printing technology is widely appreciated by consumers and small business owners, but AMT would be of paramount interest to manufacturing companies.

GE and its partner, Snecma, a French company, will use AMT or the 3D printing technology for the mass production of jet engine nozzles. GE states that each jet engine requires 10–20 nozzles and the company needs to make 25,000 nozzles annually within the next 3 years. GE managers claim that the use of these technologies improves product surface finish and quality control, cuts down production costs, provides tight tolerances, and improves part reliability. In addition, the GE additive manufacturing process uses fewer materials over conventional techniques, makes parts lighter, and eliminates scrapped material. Note that additive technology allows the parts to be built from a bed of cobalt–chromium powder. A computer-controlled laser shoots pinpoint beams into bed to melt the metallic alloy in the desired areas, creating 20 μm-thick layers one by one. Note that the additive manufacturing process is faster in making complex shapes because the machines can run around the clock without any attendant.

The GE Power and Water division that manufactures large gas, steam, wind, and water turbines has already identified parts that can be manufactured for these turbines using the additive process. In addition, the GE Healthcare division has developed a method to print transducers using the latest 3D technology, the most expensive ceramic probes used in its ultrasound machines. GE's chief technology officer says that the company is fundamentally changing the way we think about manufacturing.

Currently, GE engineers are exploring the use of MAM and 3D printing technology with a wider range of metal alloys, including some materials particularly designed for 3D printing technology. The GE aviation division is looking to use titanium, aluminum, and nickel–chromium alloys for specific applications. A single part could be made using multiple alloys allowing the part designers to tailor its material characteristics in a way not possible with conventional casting technique. For example, a blade for a jet engine or steam turbine could be made from different metals and alloys such that one end of the blade is optimized for high mechanical strength and the other end for extreme thermal resistance. GE material scientists believe that the future growth of the additive industry is strictly dependent upon the availability of better-quality materials and advanced thermoplastic materials and resins. Rapid progress in prototyping and in the application of additive processes is responsible for sustaining high manufacturing capability for critical industrial machine components. Aerospace companies manufacture massive plastic and resin-based parts. Note that the thermal and fast-setting temperature properties are best suited for aerospace applications, but AMT is still needed for certain plastic parts. The recent introduction of FDM that is made from resin promises several benefits to customers in a wide range of industries. These are examples of how high-performance polymers are being used in additive manufacturing procedures.

Technology experts believe that the issue of certain delicate part reproducibility can pose problems in the implementation of additive processes. To eliminate these problems, the current focus must be on the in-process measurement of layer temperatures, cooling rates, and other factors that control the performance and reproducibility of metal parts. It will be of great importance to point out that the industrial size and the speed of the 3D printers will allow handling of large parts needed for products such as automobiles, battlefield tanks, aircraft, and other large industrial equipment.

Summary

This chapter is dedicated to the survivability, security, and safety of the UAV. Since the aircraft is fully equipped with high-performance and costly EM and EO sensors, its survivability and security are of paramount importance. This UAV must be managed and controlled by the experienced fighter/bomber pilots sitting in the ground control stations. These ground control station operators must have comprehensive, multiyear experience as fighter/bomber pilots and should be thoroughly familiar with all aspects of the EM and EO sensors aboard the vehicle. Critical safety and security issues are discussed in great detail. For deep penetration attack and reconnaissance missions, both the RF and IR signatures must be reduced to a minimum to avoid the enemy's long-range radar detection and IR missile lock-on.

Note that these two sources present the greatest threats to the UAV platform.

Research studies undertaken by the author seem to indicate that all major issues related to UAV survivability and security must be addressed. The research studies further indicate that survivability of the UAV is strictly dependent on the flight parameters, vehicle stealth features, and IR and RF signatures. RF and IR reduction techniques have been described as thoroughly as possible. RAMs and RAPs are evaluated in the context of RF reduction technology. The benefits of RF signature reduction from RAM thickness, radiation paint, and paint thickness are summarized in the context of RF signature reduction.

IR signature reduction is of critical importance and must be given serious consideration. IR signatures are generated by aircraft structures, control surfaces, fuselage skin surface, pod-mounted sensors or weapon systems, and hot exhaust gases from the jet engine tail pipe. The IR signature may increase more than twice during afterburner flight conditions, which might attract IR missile lock-on, thereby posing a serious threat to the UAV. Exotic IR and RF signature reduction techniques are briefly discussed, which will allow the UAV to undertake deep penetration combat missions.

The hunter–killer concept for UCAVs is briefly described for undertaking deep penetration strike and reconnaissance missions in heavily defended regions using advanced EM and EO sensors and covert collected intelligence data on targets of interest. Smart optical materials are mentioned for IR optical domes and IR windows to minimize tracking errors, false targets generated by solar reflections, and pointing errors while optimizing the sensitivity of the IR spectral window. Note that smart RF and IR materials are of critical importance, when improved RF and IR performance parameters are of paramount importance. Methods to eliminate EM radiations from onboard electronic and RF devices are briefly discussed to avoid enemy detection of UAV platforms.

Requirements of advanced computer-based flight control systems are summarized for the benefit of flight control operators sitting in the ground control station with emphasis on controlling the unstable moving platform. LO technology and passive electronic counter–countermeasure techniques have been given serious considerations for the survivability of the UAV. Deployment of RAP, RAMs, and fiberglass for aircraft surfaces has been suggested for significant reduction of RCS signatures essential for UAV survivability. Other RCS reduction techniques such as shaping active loading, distributive loading technique, and passive loading concepts are briefly described to reduce UAV RCS signatures to avoid radar detection of the platform.

Various IR signature reduction techniques have been summarized to avoid IR missile lock-on. A special software, Scorpion, has been described to estimate the IR intensity level from the jet engine's tail exhaust gases as a function of aspect angle. Thermal expressions are developed to calculate the IR signature as a function of aspect angle and thrust level. The maximum IR signature appears at the tail aspect angle and the minimum at the nose aspect angle. Note that the IR signature depends on the IR intensity and the IR signature is a function of radiation intensity, thrust level, speed, aspect angle, radiation wavelength, and emissivity of the tail pipe and its nozzle materials. Note that most of the IR signatures are generated from the hot jet exhaust gases coming out of the tail section. However, some IR signatures are contributed by various aircraft surfaces but mostly from the fuselage surface. Calculated values of skin temperatures as a function of aircraft speed are provided for the benefit of readers. In addition, the percentage of total IR radiation energy over a specific IR spectral bandwidth as a function of tail pipe temperature has been calculated and summarized to illustrate the impact on IR signatures as seen by the enemy IR receiver.

MAM technology and 3D printing technique vital for tactical munitions are described in detail. These two technologies must be deployed in the design and development of pods for laser-guided and RF-guided tactical munitions, if significant reduction of RF and IR signatures is the principal requirement. As mentioned earlier, reduction of RF and IR signatures plays a critical role in the survivability, security, and safety of the UAV.

References

1. J.R. Wilson, Hunter-killer UAVs swarm battlefields, *Military and Aerospace Electronics*, July 2007, p. 24.
2. T.A. Heppenheimer, Stealth first glimpse, *Popular Science*, September 1986, pp. 74–75.

3. A.R. Jha, *Infrared Technology: Applications to Electro-optics, Photonics Devices, and IR Sensors*, John Wiley & Sons, Inc., New York, 2000, pp. 102–104.

4. Raytheon, Additive manufacturing of small tactical munitions, *Aerospace and Defense Technology*, May 2014, p. 40.

Index